ENGLISH for CARERS

by Virginia Allum

© English for Carers by Virginia Allum 2015.
Second Edition All Rights Reserved

ISBN 978-1-291-96579-7

CONTENTS

Unit 1   Introducing Yourself
Unit 2   Eating and Drinking
Unit 3   Going to the Toilet
Unit 4   Mobility and the Elderly
Unit 5   Pressure Area Care
Unit 6   Preventing Falls
Unit 7   Confusion in the Elderly
Unit 8   Giving out Medication
Unit 9   Pain in the Elderly
Unit 10   End of Life Care

Unit 1: Introducing yourself

When you talk to elderly residents make sure that you use their full name not their first name. For example: *Good Morning, Mr Browne.* If you use their first name without permission, it may sound disrespectful. It may also sound as if you are treating them like a child. Carers usually use their first name when introducing themselves. For example: *Hello, I'm Sasha. I'm one of the carers.*

Activity 1: Put the following greetings under the correct headings.

Hi, George. How are you today?

Hello, I'm Jane.

My name is Jane.

Pleased to meet you, Mrs Perry.

Good Morning, Mrs Peters.

Hi, Susan. How are things?

Nice to meet you, Maria.

Formal greeting	**Informal greeting**

> **Think about it!**
> When would it be OK to use informal greetings when you talk to a resident?

Activity 2: Watch the video on www.youtube.com/watch?v=94D-qP6qffw and answer the questions that follow.

1. The name of the carer is _____.

a) Susan

b) Jeanette

c) Jane

2. Jane introduces herself to _____.

a) the daughter of one of the residents

b) a new resident called Mrs Peters

c) a staff member

3. Mrs Peters is sitting _____.

a) on the other side of the table from Mrs Perry

b) on the right of Mr French

c) next to Mrs Perry

4. Mr French asks Mrs Peters to _____

a be more friendly

b tell him her name

c call him by his first name, George

Activity 3: Put the dialogue notes in the correct order.

1. What is your name?

2. This is Mrs Perry on your left.

3. Good Morning, Mrs Peters.

4. Please call me by my first name.

5. Pleased to meet you.

6. I'm one of the carers here.

Activity 4: Look at the picture of the video and complete the dialogue. Then watch the video again to check your answers.

Carer: _____, Mrs Peters.

Mrs Peters: Oh. Good Morning, dear. What is your name?

Carer: My _____. I'm one of the carers here. You're new here, aren't you? I'll introduce you to the people at your table.

Mrs Peters: Thank you. I don't know anyone yet.

Carer: That's all right. I know you arrived yesterday. Mrs Peters,

_____Mrs Perry on your left on the other side of the table.

Mrs Peters: Hello Mrs Perry. My _____ Maria.

Mrs Perry: Pleased to _____. My name is Susan.

Carer: And Mr French is on your left, next to you. Mr French, this is _____.

Mr French: Hello. Nice to meet you. Please call me by _____. I'm George. We are all very friendly here!

Mrs Peters: Thank you, George. You certainly are all very nice.

Activity 5: Match the beginnings and the endings.

1. What is	a) on your left.
2. I'll introduce you to	b) meet you.
3. Mrs Peters, this is Mrs Perry	c) your name?
4. Pleased to	d) first name.
5. Please call me by my	e) the people at your table.

Activity 6: Read the short article and answer the questions.

'Please don't call me dearie'

Some elderly residents are fed up with being talked to without respect. Some older people don't want carers to call them 'dear' or 'love'. They say that it makes them feel as though carers think they are children. Their grandchildren don't talk to them that way so why should carers, they say.

Carers are taught to use residents' full names when they talk to them. They use Mister, Missus and Miss with the resident's family name, never the resident's first name. But some residents say that they like carers to use terms of endearment like dear or sweetheart. They say it sounds more friendly and it makes them feel at home.

1. Some elderly residents are *pleased / annoyed* when carers talk to them disrespectfully.
2. When carers call residents 'dear' or 'love' residents *feel like their grandchildren/ feel they are treated like children*.
3. Carers should use residents' *full name / nickname* when they talk to them.
4. Words like 'dear' and 'love' are sometimes used to sound *more / less* friendly.

Places in the nursing home

Activity 7: <u>Underline</u> the stressed symbol. The first one is done for you.

<u>bath</u>room

dining room

front desk

kitchen

laundry

linen room

sitting room

treatment room

Activity 8: What happens in the rooms in the nursing home? Use the names of the rooms from activity 7 to help you.

1. Medicines and dressings are kept in the _____ room.

2. Clothes are washed in the _____.

3. Residents eat their dinner in the _____ room.

4. You can find clean sheets and blankets in the _____ room.

5. Residents can watch TV and chat in the _____ room.

6. The secretary of the nursing home is at the _____.

7. Residents wash themselves in the _____.

8. The cook prepares meals in the _____.

Giving directions

Activity 9: Watch the short video on YouTube at www.youtube.com/watch?v=XcPHcbm0ttc **and complete the dialogue. Then, practise the dialogue with a partner.**

Mrs Peters: Hello, Jane. My daughter is here to visit me. This is Sarah, my daughter.

Jane: Pleased to meet you.

Sarah: Nice to meet you too, Jane.

Mrs Peters: Where is the _____? I would like to make a cup of tea for my daughter.

Jane: That's nice. The kitchen is next to the _____.

Sarah: Jane, where is the _____? I need to wash my hands.

Jane: It's on your right when you go out of the _____.

Sarah: Thank you. Where is the sitting room? I have some cake to have with our tea.

Jane: The _____ is over there on your left. There are two chairs by the window where you can sit and have your tea.

Mrs Peters: Thank you, Jane. That sounds lovely.

Unit 2: Eating and Drinking.

It is very important to encourage the elderly to eat and drink enough each day so that they have enough energy for daily activities. Having a meal with other residents is a social activity as well so many nursing homes make sure that there is a pleasant dining room where residents can sit and chat over a meal.

Activity 1: Match the items of food and drink which might be served at lunch. Notice that we use particular expressions with each type of food. The first one is done for you.

1. A plate of	a) soup
2. A cup of	b) yoghurt
3. A bowl of	c) bread
4. A dish of	d) cake
5. A glass of	e) sandwiches
6. a slice of	f) ice-cream
7. A piece of	g) juice
8. A carton of	h) tea

(1 → e)

Watch the Youtube video at

www.youtube.com/watch?v=4LoOtS53_-M

Activity 2: Answer the questions relating to the video.

1. Mrs Peters likes to drink _____.

a) weak tea

b) black tea

c) strong tea

2. The carer is going to add _____ to the tea.

a) a slice of lemon

b) a dash of milk

c) some skimmed milk

3. Mrs Peters likes _____ of sugar in her tea.

a) two spoons

b) a teaspoon

c) half a spoon

4. Mrs Peters would also like a _____ with her tea.

a) an apple

b) a piece of cake

c) a biscuit

Activity 3: Watch the video and put the dialogue in order.

a) dash of milk

b) biscuit from the plate

c) tea weak or strong?

d) half a teaspoon of sugar

e) a cup of tea?

Expressions for 'How do you like your tea?'

If you ask how a person likes their tea and they answer, 'As it comes', they mean that they don't mind if the tea is weak or strong.

1. How many spoons of sugar do you have?

2. I like a dash of milk.

3. Can I have a slice of lemon, please?

4. Please leave the teabag in. I like it strong.

5. I like it weak.

Activity 4: Complete the short dialogues using the words below.

ice-cream   tea   cake   soup
yoghurt   juice   biscuits   bread

Carer: Would you like a bowl of (1) _____ for lunch?
Resident: That would be nice. Can I have a slice of (2)_____ too, please?

Carer: Would you like a cup of (3)_____ with your sandwich?
Resident: No, thank you. Can I have a glass of (4)_____, please?

Carer: Would you like a dish of (5) _____ after your sandwich?
Resident: No, I don't think so. I'd prefer a carton of (6)_____.

Carer: Would you like a piece of (7)_____?
Resident: No, thank you. My daughter is bringing me some (8) _____ later.

Read the short article and answer the questions true or false.

Meal times for older people

Sometimes when people get older they lose interest in some types of food. Their sense of smell may change so that tasty food aromas are not noticed. They may also have problems tasting certain food so they don't enjoy it as they used to.

Dentures which fit badly or missing teeth can make chewing difficult. They may also have less saliva in their mouths so it is more difficult to eat.

Many elderly people don't like eating large meals as they say that they don't use much energy during the day so they don't need them. Unfortunately, they sometimes get into a habit of not eating at all or eating very small meals. When this happens, they can become ill and very tired.

Activity 5: Answer the questions True or false.

1. Elderly people sometimes lose interest in eating meals. T / F

2. All elderly people stop being able to smell food. T / F.

3. Chewing is difficult if dentures do not fit properly. T / F.

4. Saliva helps you eat your food. T / F.

5. Some elderly people don't have enough energy to eat. T / F

Meal times

Activity 6: Say the words and underline the stressed syllables. Did you notice? 'breakfast' sounds like 'brekfəst'

breakfast	break - fast
morning tea	morn - ing tea
lunch	lunch
afternoon tea	aft- er - noon tea
dinner	din - ner
supper	sup - per

Activity 7: Match meal times with their correct meanings.

1. breakfast	a) a mid-afternoon snack
2. morning tea	b) a snack before bed-time
3. lunch	c) the evening meal
4. afternoon tea	d) the first meal after you wake up
5. dinner	e) a short break before lunch
6. supper	f) meal in the middle of the day

Talking about meal times

Watch the short video at

www.youtube.com/watch?v=pH9ZkkxG7Wo

Activity 8: Answer .True or False.

1. The carer says that it *was / is* time for breakfast now.
2. Mr Georgio hasn't *have / had* his breakfast yet.
3. The carer *has gone / is going* to take Mr Georgio to the dining room.
4. Mr Georgio *thinks / has thought* that it is lunch-time.
5. Mr Georgio said he *feel / feels* like a cup of tea.

Activity 9: Watch the video again and match the sentences.

1. It's time	a) your breakfast yet.
2. I've already had	b) in the morning.
3. You haven't had	c) that it was lunch-time.
4. It's 8 o'clock	d) for breakfast now
5. I really thought	e) like a cup of tea.
6. I feel	f) my breakfast.

Activity 10: Complete the short dialogues.

Mr Georgio: Breakfast? I've already had my (1)_____.

Jane: Not today, Mr Georgio. You haven't had your breakfast yet (2)_____.

Jane: It's 8 o'clock in the (3)_____. I've come to take you to the dining room for breakfast.

Mr Georgio: Are you sure it's only 8 o'clock? I would prefer to have my (4)_____ now.

Mr Georgio: Oh, well I suppose it's all right. I feel like a cup of (5)_____. Is there any tea ready?

Jane: I'll go and have a (6)_____ for you right now.

Unit 3: Going to the toilet.

Some residents need help with toileting because they have poor mobility and can't get to the toilet quickly enough. Others need help with their clothing because they don't have enough strength in their hands.

Activity 1: Match the terms with their correct meanings.

1. toilet	a) soft material which absorbs urine and faeces
2. commode	b) disposable pants used to keep an incontinence pad in place
3. bedpan	c) narrow bedpan used if a person can only lift their buttocks a little
4. slipper pan	d) portable toileting receptacle for bedbound residents
5. pad	e) chair which contains a bedpan and is used as a toilet
6. net pants	f) fixed toileting receptacle with a flushing mechanism to clean it

Activity 2: Watch the video at

www.youtube.com/watch?v=lsGFQv07tLY and answer the questions in full sentences.

1. What does Mrs Greene want to do? (go / toilet)

Mrs Greene wants _____.

2. Why does Mrs Greene need help? (not / manage herself)

Mrs Greene _____.

3. What does Mrs Greene ask the carer to do first? (help / chair)

Mrs Greene _____.

4. What does the carer ask Mrs Greene to do? (stand up)

The carer _____.

5. Does Mrs Greene want to go to the dining room after she goes to the toilet? (not want)

Mrs Greene _____.

6. What does the carer tell Mrs Greene to do? (have /drink)

The carer _____.

Activity 3: Match the beginnings and endings.

1. Will you take me	a) I'll help you up.
2. Will you help me out	b) glass of apple juice
3. Give me your arm and	c) the dining room today.
4. No, I won't go to	d) to the toilet?
5. I'll bring you a	e) on your locker.
6. I'll put it	f) of the chair first?

Activity 4: Put the notes in the correct order.

a) need help

b) give me your arm

c) take me to the toilet

d) help me out of the chair

e) have a glass of juice

f) go to the dining room

Using everyday expressions for bodily functions

Some residents are comfortable using everyday expressions for bodily functions, however, it is important not to use terms which are vulgar or rude.

It is common for people to use the expression 'go to the toilet' to mean 'have a bowel movement' but some people use this expression to mean 'pass urine'.

Formal term	Everyday term	Vulgar term
pass urine	do a wee	do a piss
have a bowel movement open your bowels	do a pooh go to the toilet	do a shit
pass flatus	pass wind	fart
be constipated	be bunged up pass hard pooh	
have diarrhoea	have loose stools have a loose bowel movement	have the runs have the trots
urine	wee	piss
faeces stool	pooh	shit

Activity 5: Complete the questions using the words below.
pooh   wee   bowels   toilet   stools

1. Have you opened your _____ today, Mrs Jones?

2. Mr Denning, have you been to the_____ today?

3. Do you want to go to the toilet to do a _____ (urine)?

4. How long have you had loose _____ (diarrhoea)?

5. What is your _____ like? Are you still constipated?

Read the short article and answer the questions true or false.
Constipation and the elderly

There are several reasons why the elderly become constipated. One cause of constipation is the lack of exercise. Older people who can't walk around easily often have sluggish bowels. They may not eat enough or eat too little fibre in their diet.

Some elderly people do not have enough to drink during the day because they worry about being incontinent of urine. Unfortunately, a low fluid intake can contribute to constipation.

Medications such as antacids and painkillers which contain codeine may also contribute to constipation. These medications make the intestines move more slowly. Diuretics or 'water pills' take water out of the intestines so it is more difficult to have a bowel movement.

Activity 6: Answer the questions about the article.

1. Elderly people may become constipated if they _____.
a) walk slowly
b) exercise too much
c) find it difficult to move around

2. Constipation may also result from eating _____.
a) too little fibre
b) fibrous vegetables
c) a diet low in fibre or not eating enough food

3. Some older people do not drink enough fluids because ___.
a) they worry they will wet themselves
b) they don't like water
c) they don't think they need a lot of water

4. Medications such as antacids and codeine can _____.
a) make a person's walk slowly
b) cause the intestine to move slowly
c) take water out of the intestines

5. Diuretics are called 'water pills' because they _____.
a) make you drink a lot of water
b) cause you to pass a lot of urine
c) make water cleaner

Bowel Chart

Look at Mr Walker's bowel chart below:

Rosemont Nursing Home Bowel Chart Month: June 2014					
Name	Jim Walker				
Room	23				
GP	Dr Sangha				
Medication	Codeine 60 mg twice a day for pain				
Bowel Regime	Laxative: Senna tablets every night Suppository: Glycerine suppository if constipated				
Bowel movments	BO - bowels opened BNO - bowels not opened				
Type of bowel movement	L loose bowel movement, diarrhoea N normal, formed stool H hard lumps, constipated				
Date	Morning	Evening	Night	Type	Comments
1/6	BO	BNO	BNO	N	Given laxative
2/6	BO	BNO	BNO	H	Given a laxative
3/6	BNO	BNO	BO	H	Given a suppository
4/6	BO	BO	BNO	N	

Activity 7: Complete the information from the bowel chart.

constipated   rectum   bowel chart   loose   medicine

1. Carers record residents' daily bowel movements in a _____.

2. A laxative is a type of _____ which makes faeces soft and easy to pass.

3. Suppositories are small, pellets shaped like bullets which are put into the _____ (bottom) to soften the stool.

4. Diarrhoea is the passing of _____, watery bowel movements.

5. A person who is _____ passes small, hard lumps of faeces or cannot have a bowel movement at all.

Talking about constipation

Watch the short video at

www.youtube.com/watch?v=srygaPLmotg

Activity 8: Match the beginnings and endings.

1. Did you go	a) a bit more water as well.
2. What was	b) some painkillers lately, haven't you?
3. You take a laxative	c) to the toilet tonight?
4. You've been taking	d) some suppositories tonight.
5. I can give you	e) every night, don't you?
6. Try to drink	f) the bowel movement like?

Activity 9: Put the dialogue in the correct order.

Resident: Yes. I just passed a few hard lumps and it hurt a lot.

Resident: Thank you. It is very uncomfortable.

Carer: I see. Are you constipated?

Resident: All right, I'll have a glass of water now.

Carer: I'll give you a laxative to make the stool softer and easier to pass.

Resident: I had some problems with my bowels tonight.

Carer: It's important to drink a lot of water when you are constipated.

Unit 4: Mobility and the elderly.

Some elderly people find moving around difficult and may use equipment like walking sticks or walking frames to help them. The main reasons for mobility impairment are being inactive, obesity and diseases such as diabetes or arthritis.

Activity 1: Match the names of equipment with the pictures.

walking frame   mobility scooter   walking stick

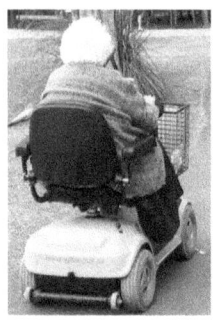　　

1.   **2.**   **3.**

Activity 2: Complete the information about the equipment above. Use the words below to help you.

mobility   arthritis   vehicle   balance   steady

1. People who are have poor _____ sometimes use a walking stick when they walk.

2. Elderly people who have _____ can use a mobility scooter to travel around.

3. A walking frame helps people to _____ themselves when they walk.

4. Elderly people who have problems walking around need to use _____ aids like walking sticks or walking frames.

5. A mobility scooter is a type of small motor _____ which people ride on if they can't walk for long distances.

Activity 3: Watch the youtube video at www.youtube.com/watch?v=ImuA-b6rmKg

1. The carer wants Mrs Peters to _____.
a) have a cup of tea
b) go for a walk in the garden
c) go to the dining room

2. Mrs Peters doesn't think she can _____.
a) go for a walk today
b) get out of the chair
c) walk very far

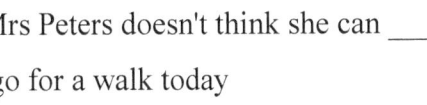

3. Mrs Peters uses _____ when she gets around.
a) a walking frame
b) a wheelchair
c) a walking stick

4. The carer offers to rub some _____ cream into Mrs Peter's knee.
a) antiseptic
b) anti-inflammatory
c) painkiller

5. The carer _____ Mrs Peters out of the chair.
a) puts
b) places
c) helps

6. Mrs Peters says she wants to _____.
a) see the roses in the garden
b) pick some roses in the garden
c) plant some roses in the garden

Activity 4: Put the dialogue in order.

a) are you steady?

b) see roses in garden

c) .use a walking stick?

d) anti-inflammatory cream on knee?

e) walk in the garden

f) can't walk far

g) help out of chair?

h) took a painkiller

Activity 5: Complete the excerpts using the verbs below.

rub   use   stand up   walk around   took   take

Carer: Do you use a walking stick when you (1) _____?

Mrs Peters: Yes. It's over there, on the chair.

Carer: Would you like me to (2) _____ rub some anti-inflammatory cream into your knee first?

Mrs Peters: No, thank you. I (3) ___ a painkiller half an hour ago. I think I'll be all right.

Carer: I'll help you out of the chair now so you can (4)_____.

Mrs Peters: If you help me to stand up, I'll (5) _____ the walking stick to steady myself.

Carer: Are you steady?

Mrs Peters: Yes. I'm quite steady now.

Carer: Can you (6) _____ my arm and we'll go for a short walk outside?

Read the short article and answer the questions.

How to use a walking frame

It is very important to have the walking frame set at the correct height. The height of the handgrips should be at the level of the wrist bone when the elbows are very slightly bent. This angle is the best position for weight bearing.

If the frame is too high, it is difficult to straighten out the elbows and the walking frame will not take enough body weight through the arms.

If the walking frame is too low, the person using it will be bent over in a poor posture. Sometimes physiotherapists set up a walking frame at a low height for people who tend to fall backwards as the lower height will encourage them to lean forwards.

The walking frame should be lifted and moved slightly in front of the person. The person leans on the walking frame and holds the handgrips. Then they take two steps of equal length into the centre of the frame.

Activity 6: Answer the questions True or false.

1. Walking frames are always set at a low height. T / F

2. The elbows should be bent at a slight angle to hold the handgrips. T/F.

3. You can't straighten your elbows if the frame is set too high. T/F.

4. You may bend over too much if the walking frame is too low. T/F.

5. If the frame is too low you might fall forward. T / F

6. To use the walking frame properly, place the walking frame behind you and take two steps into the frame. T/F

Activity 7: Match the opposite movements. The first one is done for you.

1. lean forward	a) sit down
2. lift up	b) turn around
3. sit up	c) lean back
4. stand up	d) stand in front of
5. stand up facing me	e) lie down
6. stand behind	f) put down

Activity 8: Watch the video at

www.youtube.com/watch?v=A_jfGeDbDLg

1. The carer brings a _____ for Mr George?

a) walking frame

b) walking stick

c) lunch box

2. Mr George stands up _____ the carer.

a) behind

b) at the face of

c) in front of

3. It's important to _____ your elbows slightly when you hold onto a walking frame.

a) straighten
b) hold
c) bend

4. Mr George should try not to _____ or she may fall.

a) lean forward
b) lean back
c) stand

5. The last thing Mr George does is to _____.

a) lie down
b) turn over
c) sit down

Activity 9: Unjumble what the carer says to Mr George.

1. facing / Can you / me? / stand up

2. Can you / elbows slightly? / bend your

3. put your / Can you / in front of you?/ walking frame

4. or you / Try not to / might fall / lean back

5. and sit / Can you / turn around / down?

6. Try to / on / sit down / the chair.

Activity 10: Watch again and complete the dialogue.

Carer: Yes. It's lunch time now. I've got your walking frame for you. I'll put it in front of you.

Mr George: Yes, thank you. I (1) _____ the walking frame to walk. I'm quite unsteady on my feet.

Carer: Can you stand up and (2) _____ the walking frame facing me?

Mr George: Yes. I'm a bit slow standing up, I'm afraid. It's my arthritis.

Carer: That's all right. Take your time.

Mr George: Yes, I'll have to (3) _____ my time. My joints are a bit stiff today. There we are.

Carer: That's good. Can you (4) _____ your elbows slightly and hold onto the handgrips?

Mr George: Yes. I'm ready now.

Carer: Can you (5) _____ your frame a little in front of you and take a few steps towards me?

Mr George: Ooh, it is hard to get going this morning. I'll try to start walking now.

Carer: Mr George, try not to (6) _____ or you might fall. It's better to (7) _____ a little.

Mr George: Oh. OK. Is that better?

Carer: Yes, that's better. Here's your chair now. Can you (8) _____ ? That's it. Try to sit down on the chair. It's right behind you.

Mr George: Can I sit now? I won't fall?

Carer: Yes. You can sit now.

Unit 5: Pressure Area Care

Pressure ulcers are sores which can form if there is too much pressure on a part of the body. This can happen if residents are not able to move around or change their position in bed or in a chair. The pressure on the skin blocks small blood vessels called 'capillaries' so that oxygen can't get to the tissues of the body. If the pressure is not relieved by changing position or walking around the body tissues can die causing a pressure ulcer to form.

Activity 1: Match the terms with their correct meanings

1. pressure	a) a group of cells in the body which perform a particular function
2. pressure ulcer	b) a tube-shaped channel which carries blood around the body
3. capillary	c) a wound caused by pressure on a part of the body e.g. over the hip bone
4. blood vessel	d) tiny blood vessels which join arterioles and venules
5. tissue	e) a force which is placed on someone or something

Activity 2: Watch the video at www.youtube.com/watch?v=D84FyIl38h8 and answer the questions. Use the prompts in the brackets to help you.

Activity 2:

1. What are the carers going to do? (turn/Mrs Summers/on side)
The carers are _____ turn Mrs Summers
_____.

2. Why do they need to turn Mrs Summers? (check / hip / red)
The carers need to _____ Mrs Summer's _____ because it was _____.

3. What will the carers do before dinner? (roll/onto back again/sit up for dinner)

They will _____ Mrs Summers onto _____ and _____ her up _____ .

4. Does Mrs Summers have any broken areas? (no/skin over hip not broken/red)

No, Mrs Summers does not _____ but the skin _____ her hip is red.

5. What does the carer put on Mrs Summers' bedside table? (puts drink, magazine, box of tissues / within reach)

The carer _____ Mrs Summers' _____ and _____ within Mrs Summers' reach.

Activity 3: Put the notes in the correct order.

a) sit up for dinner

b) important to change position

c) turn on side for a while

d) bedside table on other side

e) check hip

f) skin red not broken

Activity 4: Match the beginnings and endings.

1. I am going to turn you	a) a little red last time.
2. It's important to change your position	b) if you change your position.
3. I need to check your hip because it was	c) cause a nasty sore to form.
4. A small tear in the skin can	d) on your side for a while.
5. It takes the pressure off your skin	e) so you can reach it.
6. I'll put your bedside table on the other side	f) when you are in bed.

Read the short article and answer the questions which follow.

The development of a pressure ulcer

Pressure ulcers develop more easily in people who have diabetes or poor blood circulation. Poor circulation may be caused by having low blood pressure so that blood does not flow easily in the body. It may also be caused by a blockage in a blood vessel, for example during a stroke. If blood does not circulate well in the body, oxygen is not taken to the tissues and the tissues may die.

People who have very dry or over-moist skin are also more likely to develop a pressure ulcer. Residents who are incontinent of urine are at risk of pressure ulcers because their skin breaks down very easily. If residents are also bedbound or chair bound, the risk increases even more unless pressure on the body is relieved regularly. Mobility aids should be used when lifting these residents to avoid shearing forces. Shearing forces occur if residents are 'dragged' along the bed and which can tear the skin and start an ulcer forming.

Activity 5: Answer the following questions.

1. A pressure ulcer may be caused by _____.
   (a) high blood pressure
   (b) poor blood supply
   (c) blocking an ulcer with pressure

2. Oxygen is needed by _____.
   (a) the tissues of the body
   (b) the circulation of the body
   (c) the blood pressure

3. Pressure ulcers can develop if residents have _____
   (a) thin skin
   (b) broken skin
   (c) very dry or very moist skin

4. Residents who are bedbound or chairbound _____.
   (a) use a walking frame to move around
   (b) can't get out of bed or a wheelchair
   (c) prefer to rest in bed or a wheelchair as often as possible

5. Mobility aids are used to _____

(a) drag residents along the bed

(b) lift residents to avoid shearing injuries

(c) lift and drag residents in a wheelchair

Activity 6: Watch the video at www.youtube.com/watch?v=udLUwCrpB1Q and answer the questions.

1. The nurse is talking to the carer about Mrs Summers' _____.

a) Watersmith Chart
b) Waterlow Chart
c) Water Chart

2. The numbers in the chart relate to information about _____.

a) the resident's physical condition and age

b) the resident's psychological state

c) the resident's date of birth

3. It's important to check _____

a) the condition of the skin

b) the size of the bed

c) the area of the room

4. The risk of getting a pressure ulcer means _____

a) how many times a resident has had a pressure ulcer

b) how much pressure is placed on the resident

c) how likely it is that a resident will get a pressure ulcer

5. The nurse _____ the numbers in the chart to work out the level of risk.

a) adds

b) subtracts

c) divides

Activity 7: Watch the video again and complete Mrs Summers' Waterlow Chart below. Put a circle around the numbers as you hear them.

Waterlow Pressure Area Assessment

Name: Mrs Vera Summers   Aged 86   Room 16

Month: June 2014

Circle each score that is appropriate.

Build		Skin		Sex		Appetite	
Average	0	Healthy	0	Male	1	Average	0
Above average	1	Dry	1	Female	2	Poor	1
Obese	2	Red /unbroken	2	65-74	3	Fluids only	2
Below average	3	Broken	3	75-80	4	Not eating	3
				81 +	5		
Continence		**Mobility**		**Medication**		**Special Risks**	
Continent	0	Fully	0	Steroids		Poor circulation	5
Incontinent (urine)	1	Restless	1	Anti-		Smoking	1
Incont. (faeces)	2	Uses walking aid	3	inflammatories		Diabetes	4
Incont. (both)	3	Bed/chair bound	5	**4 points for either**		Stroke	4
RISK		Low 10-15		Medium 15-20		High 20+	

Activity 8: Complete the information from the Waterlow chart.

bedbound   score   medications   turned   skin

1. The _____ in a Waterlow chart is used to work out the risk of getting a pressure ulcer.
2. Carers need to check Mrs Summers' _____ frequently.
3. Residents who are _____ are at a high risk of getting a pressure ulcer.
4. If residents take _____ such as steroids, they score more on the Waterlow chart.
5. Residents who have a high risk of pressure ulcers should be _____ regularly to relieve the pressure on their bodies.

Activity 9: Match the beginnings and endings from the video.

1. I need to	a) what her skin is like?
2. I weighed	b) a smoker.
3. Can you tell me	c) drug chart.
4. She wears a pad	d) update her details.
5. I'll just check her	e) during the day and at night.
6. She's not	f) her this morning.

Activity 10: Complete the short dialogue. Use the words below to help you.

smoke   healthy   weigh   incontinence
steroids   update   above average   four

Nurse: Hello, Peter. I need to (1) _____ the details on Mr George's Waterlow chart.

Carer: Sure.

Nurse: Firstly, his build. What does he (2) _____ now?

Carer: Ah, let me see. His weight was 92 kg this morning. His build is (3) _____.

Nurse: OK. What about his skin?

Carer: His skin is (4) _____. He doesn't have any broken areas of skin. So, zero.

Nurse: Is he continent?

Carer: Yes, he is. He doesn't need (5) _____ pads at all.

Nurse: I know that he takes (6) _____ so that's....

Carer: That's (7) _____, right?

Nurse: Finally, he doesn't have diabetes but he is a (8) _____.

Carer: That's right. He still smokes a bit.

Unit 6: Preventing Falls

Falls can be a serious problem in hospitals and nursing homes. Falls can cause very serious injuries like broken hips and even death. They can lead to the need for a long hospital stay and they also reduce mobility so that residents depend more on carers for their care. For these reasons, it is very important for carers to prevent falls as much as possible.

Activity 1: Match the terms with their correct meanings

1. to fall	a) to slide on some liquid and fall onto the floor
2. to slip	b) to stop something from happening
3. to trip	c) to hurt yourself
4. to injure yourself	d) to lose your balance and collapse onto the floor
5. to prevent	e) to fall over an object and hurt yourself

Activity 2: Watch the video at www.youtube.com/watch?v=apZ8Syyh0FI and answer the questions.

Activity 2:

1. What happened to Mrs Mitchell?
a) She falls on the floor.
b) She fell on the floor.
c) She has a fall on the floor.

2. Why did Mrs Mitchell have the accident?
a) She slipped on some water on the floor.
b) She tripped over her bag on the floor.
c) She fell over when she tried to go to the toilet.

3. Why does she have to go to the toilet a lot?

a) She takes tablets which make her pass a lot of urine.

b) She forgets that she has already been to the toilet.

c) She has an infection and has to go to the toilet every hour.

4. Why is Mrs Mitchell upset?

a) She thinks the carer will be angry with her.

b) She is annoyed that the carer was too busy to help her.

c) She has been incontinent and is embarrassed about it.

5. Why does the carer have to wait for the nurse to come?

a) So the nurse can help her get Mrs Mitchell to stand up.

b) So that the nurse can check that Mrs Mitchell hasn't injured herself.

c) So the nurse can take Mrs Mitchell to the toilet.

Activity 3: Match the sentences below.

1. I fell over when	a) because I wet myself.
2. I am very embarrassed	b) because you are very busy.
3. I didn't want to bother you	c) I've hurt myself.
4. I don't think	d) without some help.
5. I can't get up	e) I tried to go to the toilet.

Activity 4: Complete the short dialogue using the words below.

bother   hurt myself   mean   fell over   feel   wet myself

Carer: What happened, Mr Jeffreys?

Resident: I (1) _____ when I tried to go to the toilet.

Carer: Oh dear. That's no good. Are you all right?

Resident: I am very embarrassed because I (2)_____.

Carer: I see. I know that's not very nice for you but you didn't (3) _____ to do it. Why didn't you call us to help you?

Resident: I didn't want to (4) _____ you because you are very busy.

Carer: Oh please don't think that, Mr Jeffreys. We are here to help you. Do you (5) _____ OK now?

Resident: Yes. I don't think I've (6)_____.

Carer: I'll ask the nurse to check you first then we'll help you stand up.

Resident: Thank you, Suzy. I can't get up without some help.

Grammar point: Reflexive verbs.

Did you notice the verbs 'hurt myself' and 'wet myself'?

These verbs are reflexive verbs because they describe an action which 'reflects back' on a person. They describe actions which the person does himself or herself. Other reflexive verbs which you may use are:

to dress yourself   *Can you dress yourself?*
to feed yourself   *I can feed myself - I don't need any help.*
to wash yourself   *He can't wash himself without help.*
to look after yourself   *My grandmother can't look after herself.*

Read the short article and answer the questions which follow
Preventing falls

Older people who can't see well, especially at night, may not notice objects on the floor and may trip over them. It's a good idea to keep the floor clear of clutter and avoid having rugs on the floor. Putting on a night light or keeping a light on in the bathroom is also advisable.

Many residents fall because they are take medicine like painkillers which make them feel sleepy. Elderly residents who take medication to lower their blood pressure may feel unsteady when they stand up. If they also have problems with their balance, they may fall and hurt themselves. It's important that residents have a call bell within reach at all times. They should be encouraged to call for help before they stand up or get out of bed.

A resident who can't hold their urine or faeces until they reach the toilet might slip on the wet floor. It is important to make sure that elderly residents who have incontinence problems are taken to the toilet frequently. It is a good idea for them to wear incontinence pads at night. A commode can be left next to the bed for night-time use as well.

Activity 5: Answer the following questions.
1. Residents may _____ things which are left on the floor.
(a) tripped
(b) have tripped over
(c) trip over

2. The floor should be _____.

(a) kept clean

(b) kept clear of small objects

(c) covered with a rug

3. A night light helps residents who _____.

(a) have poor eye sight especially at night

(b) want to call the carer for help

(c) like to read at night

4. Some medication can make residents feel _____.

(a) like going to bed

(b) less pain

(c) very tired or dizzy

5. Residents use a call bell to _____.

(a) ask for a carer to come

(b) ask for a cup of tea

(c) get out of bed

6. Residents who are incontinent on the floor may _____.
a) want to get to the toilet
b) slip on the urine
c) trip over a bag on the floor

7. Carers should _____ incontinent residents frequently.
a) toilet
b) wash
c) feed

8. A commode is a _____.
a) lounge chair
b) small sofa
c) chair with a removable bed pan

Look at the Fall Risk Factor Checklist and complete the activity which follows.

FALL RISK FACTOR CHECKLIST **Resident: Mr George Friend Room 1**		Yes / No
Vision	poor vision wears glasses	
Mobility	unsteady gait / poor balance uses mobility aid unsafe shoewear	
Behaviour	confused at times disoriented at night	
Continence	incontinent of urine incontinent of faeces / uses laxatives	
Medication	painkillers sleeping tablets diuretics (water tablets)	
FALL RISK	LOW MEDIUM HIGH **Important:** If resident has a HIGH falls risk, Commence Fall Risk Reduction (night light, frequent toileting, call bell within reach, keep environment clear at all times)	

Activity 6: Complete the information from the Fall Risk Factor Checklist.

toilet   drowsy   wear   surroundings
        incontinent   unsteady

1. A resident who has poor vision may need to _____ glasses.

2. If you have an _____ gait, you have poor balance.

3. A person who is disoriented is not sure of their _____.

4. Residents who take laxatives often need to get to the _____ quickly to have a bowel movement.

5. Residents who are _____ of faeces may say that they have 'soiled themselves'.

6. Sleeping tablets may make residents still feel _____ when they wake up.

Activity 7: Watch the video at www.youtube.com/watch?v=Zrhb2eQ1yZ0 and answer the questions.

1. The nurse asks the carer if Mr Field _____.
a) has tripped over today
b) is going to have a fall today
c) has had a fall today

2. Mr Field had a fall because he _____.
a) tripped over his slippers
b) couldn't see his way to the toilet
c) wouldn't ring his call buzzer

3. The night staff forgot to _____.
a) check his call buzzer
b) take Mr Field to the toilet
c) turn on the night light

4. Mr Field shouldn't wear his old slippers because _____.
a) he might fall over easily
b) they are not comfortable
c) he should wear shoes

5. The nurse asks the carer to make sure Mr Field _____.
a) leaves his call buzzer on the locker
b) uses his call buzzer to call for help
c) doesn't fall over again

Activity 8: Watch the video again and complete the Falls Risk Assessment Tool on page 5 for Mr Friend. What level of falls risk do you think he has? (Low, medium or high) What falls risk reduction strategies does the nurse suggest for Mr Friend?

Activity 9: Complete the answers to the questions.

1. Nurse: Has Mr Field had any falls today?

   Carer: Not today, but /did have/ fall last night.

2. Nurse: Was his night light put on last night?

   Carer: No / don't think / was.

3. **Nurse:** What about his call bell? Does he use it all the time?

   Carer: don't think / likes / much. He often seems / leave / on / locker.

4. **Nurse:** Could you talk to him about using his call buzzer?

   Carer: Yes / might be able / explain / important / ask for help.

Activity 10: Complete the short dialogue. Use the words below to help you.

wear   brought   switched   ring   trip over   wake up

Carer: Hello, Mr Field. I need to make sure that your night light is -_____ on.

Resident: Why do I need the light on?

Carer: It's so you can see if you _____ in the night and need to go to the toilet.

Resident: Oh, yes. That's a good idea.

Carer: Mr Friend, it would be a good idea if you didn't _____ your old slippers any more.

Resident: But they are so comfortable!

Carer: I know but you might _____ easily if you wear them.

Resident: Well, my daughter _____ me some new ones so I guess I can wear those.

Carer: Good idea. One last thing - could you use the call buzzer if you want to get up at night?

Resident: I don't like to _____ late at night.

Unit 7: Confusion in the elderly

Acute confusion in the elderly is sudden and occurs within hours or days. It is also called delirium. It is usually caused by something which can be treated so that the resident returns to their normal self. Confusion is usually a symptom of acute illness, for example urinary tract infection, but it also has its own symptoms.

The main symptoms of confusion include a change in a person's mental state. They may be hyperactive and move around a lot or lethargic and appear to be sleepy. A confused person may have trouble thinking clearly and may even hallucinate which means that they see things which are not really there.

Activity 1: Complete the sentences. Use the words below.
confused   real   serious   active   tired   acute
1. An acute illness is a _____ illness which happens quickly.
2. Delirium is another term for _____ confusion.
3. Residents who have urinary tract infections may sometimes become_____.

4. A person who is hyperactive is extremely _____ and walks around a lot.

5. Lethargy means feeling very _____ or sleepy.

6. Hallucinations are feelings that something you see is _____ when it is not.

Watch the video at www.youtube.com/watch?v=rWWBpLfj0nU and answer the questions which follow.

Activity 2:

1. Why is the carer worried about Mrs Littlewood?

a) She has become very sad.

b) She has become confused.

c) She has stopped talking to the other residents.

2. What is Mrs Littlewood's behaviour like?

a) She is quite aggressive.

b) She feels very sleepy.

c) She is relaxed and calm.

3. What else is Mrs Littlewood doing which is not normal?

a) She is having tea with her mother.

b) She is wandering out of the nursing home.

c) She thinks she sees people who are not really there.

4. Does Mrs Littlewood have any bowel problems?

a) Yes, she has constipation at times.

b) No, she has a daily bowel movement.

c) The carer is not sure about her bowel habits..

5. What does the nurse think has caused the acute confusion?

a) A change in medication.

b) A change in diet.

c) A change in her room.

Activity 3: Complete the sentences. Use the words below.

dementia   real   delirium   drugs   confused

1. People who are _____ may not recognise a person they usually know.

2. A hallucination can be something that you see, hear or smell which is not_____ .

3. Acute confusion, also called _____ can happen very suddenly.

4. Some _____ e.g. morphine can cause confusion.

5. Confusion is different from _____ because it can usually be treated.

Activity 4: Match the questions and answers.

1. Which room is she in?	a) No, she hasn't had any problems with her bowels.
2. What's the problem?	b) She keeps saying that her mother is having a cup of tea with her.
3. What about her bowels? Is she constipated?	c) The doctor changed her painkillers this week.
4. You mean, she's seeing things that aren't there?	d) She's in room 18, next to Mrs Jones.
5. Has she had a change in medication?	e) Well, she's not making sense at all.

Read the short article and answer the questions which follow.

How to manage confusion in an elderly resident

1. Communicate clearly and calmly. Try to avoid changing staff often as this can increase confusion.

2. Make sure that residents have enough to eat and drink as they may forget to have meals if they are not reminded of meal times.

3. Check that residents are not in pain and that they feel comfortable. Also ensure that residents are helped to the toilet frequently and kept clean and dry.

4. Place personal items and photographs at the bedside. Use clocks and calendars to remind residents of the time of day and the date.

5. Only use medication like sleeping tablets or sedatives if other methods have failed as these drugs may make the confusion worse.

Activity 5: Answer the following questions.

1. It's better to _____ who look after confused residents.

(a) have a mixture of nurses and carers

(b) have a small number of carers

(c) have private nurses

2. Residents should be reminded about meals or _____.

(a) they may eat too much

(b) carers have to make them a sandwich

(c) they may not eat and drink enough

3. Residents may become more confused if they _____.

(a) are not toileted frequently

(b) don't have a shower every day

(c) have diarrhoea

4. Residents may feel less confused if they_____.

(a) can tell the time

(b) have familiar objects around them

(c) know what time of year it is

5. Medications such as sedatives should be _____.

(a) given to residents if they become confused

(b) given to residents if they want to go to sleep

(c) used only if residents become extremely confused

Activity 6: Watch the video at www.youtube.com/watch?v=hjqfY0HAXFY and answer the questions.

1. Mr Browne says that he _____.

a) can't remember where he is

b) doesn't know who the carer is

c) wants to know the time

2. The carer shows Mr Browne where _____.

a) his glasses are

b) his newspaper is

c) the kitchen is

3. Mr Browne wants to _____

a) go into the garden

b) go to his friend's house

c) go to work

4. The carer tries to _____

a) convince Mr Browne that he is confused

b) help Mr Browne find the front door

c) interest Mr Browne in having a warm drink

5. Mr Browne finally _____.

a) agrees to go to the dining room

b) returns to his bedroom

c) leaves for work

Activity 7: Complete the phrases from the dialogue. Use the words below to help you.

day-time   dining   room   carers   drink   dinner

1. My name is Rosemary. I'm one of the_____.

2. It's not time to go out now. You've just had your _____ with the other residents.

3. Remember that you only work in the_____. It's time to get ready for bed soon.

4. What about having a small _____ first? You still have time.

5. That's your bag, isn't it? I'll bring it into the _____ for you.

Look at the carer's CAM assessment of Mr Browne. A CAM Assessment is the Confusion Assessment Method. Answer the questions True or False.

Resident's Name: Mr Timothy Browne Date of Birth: 9/6/1925 Person conducting the assessment: Rosemary Daintry Date of CAM assessment: 16 July, 2014			
Feature	**Questions**	**Yes**	**No**
Feature 1: **Acute Onset**	Has there been a sudden change in the resident's mental state?	√	
Feature 2: **Inattention**	Does the resident find it difficult to pay attention to what is said?	√	
Feature 3: **Disorganised thinking**	Was the resident's conversation unclear or irrelevant?	√	
Feature 4: **Altered Level of Consciousness**	How would you rate this patient's level of consciousness? Yes • Alert (normal) No • Hyperalert (overactive) √ • Lethargic (drowsy) • Stupor (difficult to arouse) • Coma (can't be aroused)		√
If Features 1 - 3 are ticked 'Yes' and Feature 4 is ticked No, a diagnosis of delirium is made.			

Activity 8:

1. The carer does the CAM assessment of Mr Browne in July.
T / F

2. Mr Browne has had a gradual change in his mental state.
T / F

3. It was difficult for Mr Browne to follow the carer's conversation. T / F

4. Mr Browne explains things clearly to the carer.
T / F

5. Mr Browne was very tired when the carer did the assessment.
T / F

Activity 9: Unjumble the sentences.

1. very / Mr Browne / confused / has become

2. attention / the carer / Mr Browne's / diverts

3. Mr Browne / which don't / says things / make sense

4. a lot / Mr Browne / lately / wanders

5. doesn't recognise / at all / Mr Browne / the carer

Activity 10: Complete the short dialogue.

a) I need my bag first.

b) I'll get you some warm milk.

c) Where's the door?

d) I don't know you.

e) I have to go to work.

Resident: Who are you? (1)_____ . Do you work here?

Carer : Yes, Mr Browne. I work here. My name is Rosemary. I'm one of the carers.

Resident: Where's the door? (2) _____

Carer: Oh but it's night-time now. It's not time to go out now. You've just had your dinner with the other residents.

Resident: No, no, no. I have to go to work now. Look.

(3) _____ Tell me where the door is now. I need to go.

Carer: What about having a small drink first? You still have time. Let's go into the dining room and (4) _____

Resident: Yes, I'm quite thirsty. (5) _____ I'll have to go out soon.

Carer: Sure. That's your bag, isn't it? I'll bring it into the dining room for you.

Unit 8: Giving out medication

Older people in nursing homes can be quite vulnerable and often rely on carers for many of their everyday needs. Elderly people who have several medical conditions and may take several medications every day. Polypharmacy or taking several medications can increase the risk of drug errors.

Residents who are mentally alert and capable of managing their own medications are usually able to self-medicate. They are responsible for taking their own regular medications. They keep their tablets in their bedside lockers.

Medications called 'as required' or 'prn medications' are medications like painkillers which are given to residents if they are needed. Residents have to ask for these medications if they need them.

Activity 1: Match the terms with their meanings.

1. polypharmacy	a) the amount of a medication you need to take
2. over-the-counter	b) a liquid medication
3. self-administration	c) medication which is only taken if you need it
4. dosage	d) taking many medications each day
5. when required	e) giving yourself your own medications
6. elixir	f) medicine bought without a prescription

Activity 2: Complete the sentences using the words below.

mls   pharmacy   drug errors   prn   milligrams   swallow

1. You need to be careful about _____ if a resident takes a lot of medications.
2. Some residents have their medication as an elixir because it is easier to _____.
3. The dosage of a medication is usually in _____ or mgs for tablets.

4. Liquid medicine is measured out in millilitres or_____ .

5. The term '__' is a Latin term which means 'whenever you need it.'

6. Over-the-counter medications can be bought at the ____ without a prescription.

Activity 3: Watch the video at www.youtube.com/watch?v=f_gpiC8fCco and answer the questions which follow.

1. Mrs Grey is able to _____.
   a) take as many painkillers as she wants
   b) buy her medication herself at the pharmacy
   c) take her regular medications herself

2. The carer checks whether Mrs Grey has taken _____.
a) her temperature
b) all her tablets
c) her afternoon medication

3. Mrs Grey understands _____.
a) what her tablets are used for
b) why the carer has to give her the tablets each morning
c) that she should ask the carer for her tablets

4. The carer explains that an elixir would be _____.
a) more expensive than a tablet
b) easier to swallow than a tablet
c) hard to take

5. The painkiller Mrs Grey asks for is a _____.
a) regular medication
b) kind of over-the-counter medication
c) type of 'as required' medication

Activity 4: Match the beginnings and endings

1. Have you taken	a) and Omeprazole 20 mg?
2. Do you mind if I check	b) one of my tablets which I am having trouble with?
3. Did you take Furosemide 40 mg	c) all your medication this morning?
4. Can I ask you about	d) painkiller this morning, please?
5. Could I have a	e) against your drug chart to make sure?

Read the article and answer the questions which follow.

The reasons for medication errors

The most common types of medication errors are incorrect crushing of tablets, not supervising the taking of medications, giving medications at the wrong time and forgetting to give medication. It's more common for nursing home staff to make an error in the morning which is frequently the busiest time of the day.

There are more mistakes made with liquid medication than with tablets.

The most common reason given for medication errors is being interrupted during the preparation or administration of medications. Some nursing homes insist that staff wear a red tabard when they are giving out medications saying that they must not be interrupted.

It's a good idea to include mentally alert residents in the medication process. Residents should be encouraged to ask questions about their medications if they think that an error has been made in the dosage or medication itself. This can prevent an error from occurring and is called a 'near miss.'

Activity 5: Answer the following questions.
1. Residents who are not watched when they are given their medications may _____.
a) miss a dose
b) refuse to take their medication
c) take too much medication

2. If medication is given at the wrong time, it is classed as a _____.

a) medication administration

b) medication error

c) type of medication

3. To stop interruptions during drug rounds, health workers wear _____.

a) a clean uniform

b) an apron when they hand out the meals

c) a type of red apron with 'Do not Disturb' written on it

4. Carers should encourage mentally alert residents to _____.

a) ask when their meal is ready

b) ask questions about their medications

c) ask for more medication

Activity 6: Watch the video at www.youtube.com/watch?v=Y-CflACfJbY and complete the sentences from the dialogue. Use the words below to help you.

poison   crush   pharmacist   tablets   permission

1. The staff had a meeting to discuss how to give Mr George his_____.

2. Mr George thinks the carer is trying to _____ him.

3. The _____ explained that Mr George should continue to take his tablets.

4. The carer has written _____ to hide Mr George's tablets in his food.

5. The nurse will give the carer a mortar and pestle to _____ the tablets.

Activity 7: Complete the dialogue using the expressions below.

a) Does he have to have the tablets?

b) How will I crush the tablet?

c) So, it's OK to do that?

d) Why did you have the meeting?

Nurse: Julie, I need to fill you in about Mr George's best interest meeting.

Carer : OK. (1) _____

Nurse: When I try to give him his blood pressure tablets he thinks I am trying to poison him.

Carer: I see. (2) _____

Nurse: Yes. It's very important that he has his blood pressure tablets. We need to hide his medication so he doesn't know he is taking it.

Carer: (3) _____

Nurse: It's OK because we have written permission and because Mr George is no longer able to make the important decision himself

Carer: (4) _____

Nurse: I'll give you a mortar and pestle to grind the tablet to a powder.

Look at Mrs Simpson's drug chart and answer the questions.

Medication	Appearance of tablet	Reason for medication	7 am	1 pm	3 pm	6 pm	11 pm
Resident's Name: Mrs Vera Simpson					DOB: 3/7/1933		
Furosemide 40mg	round white	diuretic (water tablet)	1	1			
Omeprazole 20mg	white and yellow capsule	stomach liner	1			1	
Senna	round brown	laxative				2	
Methotrexate 7.5 mg once a week on Monday only	yellow oval	Rheumatoid Arthritis	3				

Activity 8: Answer the questions about the drug chart.

1. The carer does the CAM assessment of Mr Browne in July. T/F
2. Mr Browne has had a gradual change in his mental state. T/F
3. It was difficult for Mr Browne to follow the carer's conversation. T / F
4. Mr Browne explains things clearly to the carer. T/F
5. Mr Browne was very tired when the carer did the assessment. T/F

Activity 9: Unjumble the sentences.

1. very / Mr Browne / confused / has become
2. attention / the carer / Mr Browne's / diverts
3. Mr Browne / which don't / says things / make sense
4. a lot / Mr Browne / lately / wanders
5. doesn't recognise / at all / Mr Browne / the carer

Activity 10: Complete the short dialogue with the missing sentences and questions.

a) I need my bag first.

b) I'll get you some warm milk.

c) Where's the door?

d) I don't know you.

e) I have to go to work.

Resident: Who are you? (1)_____ . Do you work here?

Carer : Yes, Mr Browne. I work here. My name is Rosemary. I'm one of the carers.

Resident: Where's the door? (2) _____

Carer: Oh but it's night-time now. It's not time to go out now. You've just had your dinner with the other residents.

Resident: No, no, no. I have to go to work now. Look. (3) _____ Tell me where the door is now. I need to go.

Carer: What about having a small drink first? You still have time. Let's go into the dining room and (4) _____

Resident: Yes, I'm quite thirsty. (5) _____ I'll have to go out soon.

Carer: Sure. That's your bag, isn't it? I'll bring it into the dining room for you.

Unit 9: Pain in the Elderly

Many elderly residents don't like asking for pain medication because they think that carers will consider them demanding. They sometimes also believe that asking for pain medication is a sign of weakness. This is because some elderly people were brought up not to complain about pain.

Sometimes the elderly are not comfortable taking pain medication because they worry that the medication is addictive. They might also be concerned about the side effects of some painkillers, for example constipation or drowsiness.

Activity 1: Match the terms with their meanings

1. painkiller	a) something which forms a habit
2. relieve pain	b) unwanted extra effect of a drug
3. anti-inflammatory gel	c) take pain away
4. addictive	d) medicine which takes away pain symptoms
5. side effect	e) non-greasy, clear cream which reduces swelling

Activity 2: Watch the video at www.youtube.com/watch?v=SiAGgfazA94 and answer the questions which follow.

1. The carer thinks Mrs Cotterall is _____ .
a) complaining about pain too much
b) in pain but doesn't want to admit it
c) late for going out in the garden

2. The pain in Mrs Cotterall's hip is _____ .
a) causing difficulties when she walks
b) stopping her from getting out of bed
c) extremely severe

3. Mrs Cotterall thinks if she tells the carer that she has pain, ____ .
a) the carer will be annoyed
b) the carer will give her some pain killers
c) the carer will think she is complaining too much

4. The carer explains that not all painkillers are _____ .
a) addictive
b) tablets
c) hard to swallow

5. Anti-inflammatory gel can be _____.
a) used instead of painkillers
b) rubbed into painful areas
c) used several times a day as it is not addictive

Activity 3: Complete the sentences using the words below.

put up with   addictive   walk around

rub   relieve   painful

1. Your hip was _____ yesterday, wasn't it?

2. You don't have to _____ the pain.

3. The pain does make it hard to_____.

4. There are some things we can give you to _____ the pain.

5. There are some painkillers that are not_____ .

6. It's an anti-inflammatory gel which I _____ into your hip.

Read the article and answer the questions which follow.

Talking about pain in the elderly

Over 80% of elderly residents in nursing homes suffer from pain, particularly if they have chronic diseases such as arthritis, cancer or joint problems. Unrelieved pain can cause health issues such as pneumonia or the formation of blood clots because pain can result in reduced mobility.

It can be difficult to assess pain in the elderly as many older people assume that pain is part of growing older. Also, they may have been brought up to believe that they shouldn't admit to having pain because it may be seen as weakness.

Unfortunately, many healthcare workers believe that elderly residents are not in pain because they don't report any pain or ask for painkillers. It is sometimes also assumed that residents with dementia no longer feel any pain. They are often not given adequate pain relief or not offered pain relief at all.

Activity 4: Match the terms with their meanings.

1. chronic	a) inflammation of the joints
2. arthritis	b) the ability to move around
3. mobility	c) condition of mental deterioration
4. joints	d) something that lasts long
5. dementia	e) part where two bones meet

Activity 5: Complete the sentences using the words below.

independent   aching   dementia   addicted   lasts

1. A chronic illness is a condition which _____ for a long time.
2. Residents who have arthritis often have _____ hips or knees.
3. Having poor mobility can make residents less_____.
4. If residents become _____ to painkillers they feel that they must take them.

    5. Some healthcare workers don't believe that residents with _____ feel any pain.

Activity 6: Watch the video at www.youtube.com/watch?v=kuoDHd_sPNE and match the beginnings and endings of the sentences.

1. Can you tell me what	a) feel like?
2. Can you tell me how	b) do the exercises the physiotherapist gave you?
3. What does the pain	c) the pain's like at the moment?
4. Are you able to	d) painkillers before you do the exercises?
5. Are you taking your	e) painful it is on the pain scale?

Activity 7: Complete the dialogue using the expressions below.

a) it's an eight.
b) I try to but I can't manage.
c) It's still very painful.
d) I don't want to take too many painkillers.
e) It's a really sharp pain

Carer: Can you tell me what the pain's like at the moment?

Resident: Yes. (1)_____. I don't think I'm ever going to be all right again.

Carer: Can you tell me how painful it is on the pain scale?

Resident: It's a seven this morning..no, actually (2) _____.

Carer: Well, if the pain stops you from moving around then it's severe pain. What does the pain feel like?

Resident: It seems to ache most of the time. (3) _____ when I try to lift my arm.

Carer: Are you able to do the exercises the physiotherapist gave you?

Resident: No, I can't do them. (4) _____. It's too painful.

Carer: Are you taking your painkillers before you do the exercises?

Resident: No. (5) _____. It's not good for you.

Activity 8: Put the notes in the correct order.

Sharp pain when lift arm

What pain like?

Painkillers due now

How painful on pain scale?

Wait 20 minutes, able to do exercises

Able to do exercises?

Take painkillers before exercise

Important not miss dose

Post-operative instructions Shoulder Surgery

* After your operation you will wear a collar and cuff sling to support your shoulder during the day. You may take it off at night.

* Do your physiotherapy exercises several times a day. Start by swinging your arm gently five times. After one week, lift your arm to a 90^0 angle. After two weeks, lift your arm to the level of your shoulder.

* It is very important that you take your painkillers regularly. Make sure that you take the tablets 15 to 30 minutes before you exercise.

* You can expect to have moderate pain for at least one week then your pain level should lessen. Please call your doctor if you have any problems.

Activity 9: Answer the questions about the post-op instructions.

1. After a shoulder operation you need to wear a sling to support your shoulder. T / F

2. A collar and cuff is a type of sling which is put around the neck like a collar. T / F

3. You must wear the sling all day and all night. T / F

4. Physiotherapists give patients exercises to help patients walk safely. T / F

5. One week after a shoulder operation you should be able to raise your arm over your head. T / F

6. You must take tablets to reduce the pain and so you can do your exercises. T / F

7. It's better to take your painkillers after you do your exercises. T / F

8. If you have less pain after your operation, you must seek medical help. T / F

Unit 10: End of Life Care

When residents move into a nursing home they usually complete an advance care plan to inform staff about their wishes at the end of life. Residents who have a progressive illness or the early stages of dementia need to agree on a plan for their care should they need palliative care or terminal care.

Palliative care is the holistic care of patients who have an advanced degenerative illness. Terminal care is provided in the last few days of life while a person is dying. Both types of end of life care include physical, psychological, social and spiritual support.

Activity 1: Match the terms with their meanings.

1. Advance Care Plan	a) something which treats the whole person
2. Palliative Care	b) describes something which gets worse and worse
3. Terminal Care	c) plan which describes care at the end of life
4. holistic	d) care which supports a patient but does not cure a disease
5. degenerative	e) care of a person in the last days or hours of life

Activity 2: Watch the video at www.youtube.com/watch?v=DjdvBCGSCPQ and answer the questions which follow.

1. The carer needs to _____.
a) update Mrs Downing's care plan.
b) change Mrs Downing's advance care plan.
c) update Mrs Downing's advance care plan.

2. An advance care plan sets out a person's _____.
a) wishes at the end of life
b) financial plans for the future
c) funeral plans.

3. Mrs Downing says she wants to have _____.
a) treatment for her condition
b) enough medication to relieve any pain at the end of life
c) an operation to cure her illness

4. The carer explains that Mrs Downing's _____.
a) family will be able to visit her but not stay at the end of life
b) friends and families can call and ask about her condition
c) relatives can stay with her at the end of life

5. The carer asks if Mrs Downing would like _____.
a) to have her favourite music playing at the end of life.
b) a photograph of her family in her room
c) to go to a music concert

6. The carer offers Mrs Downing _____ by offering to call Father Martin to come and visit her.
a) physical assessment
b) psychological care
c) spiritual care

Activity 3: Empathy is the ability to try and understand how another person feels.
Which of the sentences from the video are examples of empathy?

1. I know it must be difficult for you to talk about end of life care.

2. That's good!

3. It's quite normal to worry about that ...

4. Yes, of course it's all right.

5. Is there anything that you'd like to have in your room?

6. We can play any music you like!

Read the article and answer the questions which follow.
End of Life Care

End of life care includes support for both residents and their families. When an elderly person moves into a nursing home, an initial advance care plan is made between the resident, their doctor and the nursing home. Issues such as wishes for the end of life as well as legal documentation are discussed and noted in the resident's care plan.

Residents talk about whether they want to be resuscitated or not, if they have a cardiac arrest. The order not to resuscitate is called DNAR or 'do not attempt to resuscitate'.

Some residents also decide that they want to stop medical treatment if they feel that it is no longer needed. This is called an ADRT order or Advance Decision to Refuse Treatment.

It's very important to reassure residents that medication which helps with symptoms like pain or nausea will not be stopped so that they are kept comfortable up to the time of their death.

End of life care is holistic care as it deals with a resident's physical, emotional, psychological and spiritual needs. As a resident's condition deteriorates, the advance care plan is updated to ensure that all the resident's wishes are respected and that the resident feels supported until the time of death.

Activity 4: Complete the statements about preparing for end of life care.

heart attack   spiritual   cared for   depression   unnecessary

1. An advance care plan describes how residents want to be _____ at the end of their lives.

2. Residents often ask to stop _____ treatment at the time of death.

3. Many elderly residents do not want to be 'brought back to life' if they have a_____.

4. Examples of psychological care include talking about _____ or concerns about dying.

5. Many residents ask for _____ care from a minister of religion as it gives them comfort.

Look at Mr Freeman's Medical Care Plan Summary and choose the correct answers.

Personal Information	
Resident's Name: Peter Freeman DOB: 11/4/1923	Name of GP: Dr Amanda Love ph: 02356 213 556
Next of Kin: June de Souza (daughter) ph: 07456 556 665	Power of Attorney: June de Souza
Advance Care Planning		
DNAR (Do not attempt perform CPR)	<u>YES</u> / NO	Date: 6/7/14
ADRT form (Advance Decision to Refuse Treatment)	<u>YES</u> / NO	Date: 16/7/14
Document Review		
Nurse: *G.Sands RN (GLORIA SANDS)*	DNAR reviewed	Date: 2/8/2014
Nurse: *G.Sands RN (GLORIA SANDS)*	ADRT reviewed	Date: 2/8/2014

Activity 5:

1. Amanda Love is the name of Mr Freeman's <u>nurse / doctor.</u>

2. Mr Freeman's next of kin is his <u>daughter / daughter-in-law</u>.

3. If Mr Freeman has a cardiac arrest, he <u>does / does not want</u> to be resuscitated.

4. The ADRT form was <u>made / reviewed</u> on the 6th of July, 2014.

5. The DNAR form was reviewed by the <u>carer / nurse</u> in August.

Activity 6: Watch the video at www.youtube.com/watch?v=ENntqwC5Amc and put the notes in order.

worried about death rattle
keep talking to your father
injection stop noisy breathing
seems very agitated
injection to calm him down
Dad trouble breathing
keep changing position and doing mouth care
worried - too much medication
very small dose morphine - help with breathlessness

Activity 7: Answer the questions True / False

1. Mr Bartoli used to have trouble breathing. T / F

2. The carer tells Mr Bartoli's son that his father may still be able to hear him. T / F

3. Very small doses of morphine can help with terminal breathlessness. T / F

4. Some people have a type of noisy breathing called the 'death rattle' when they sit up. T /F

5. The nurse can give an injection to stop the noisy breathing whenever it is needed. T / F

6. The carers tell Mr Bartoli's son to come back with his sister to visit their father. T / F

Activity 8: Complete the short dialogue below.

a) give him another injection
b) keep talking to him
c) make him more comfortable
d) try not to worry

Son: Can you help my father? He's having trouble breathing.
Carer 1: Yes, sure. I can (1)_____.
Son: He seems very agitated at the moment.
Carer 1: I'll ask the nurse to (2) _____ to calm him down.
Son: The nurse said she was giving him some morphine. I'm worried about that because morphine slows your breathing.
Carer 1: Please (3) _____ because your father is only getting a very small dose of morphine. It will help his breathlessness.
Son: Is there anything else I can do to help my father?
Carer 1: Just (4) _____ even if he doesn't talk back to you. He can still hear you.

Activity 9: Match the beginnings and the endings of the sentences.

1. He's been having	a) that I'm here.
2. He's not aware	b) about the death rattle.
3. He seems	c) because of the breathlessness.
4. My sister is worried	d) a bit of trouble breathing.
5. His mouth and lips are very dry	e) very agitated.

Activity 10: You are Mr Bartoli's carer. Answer the questions Mr Bartoli's son asks you. Unjumble the dialogue.

a) He's having a very small dose to help with the breathlessness.

b) I'll give him some mouth care right now to help with the dryness.

c) Yes. The nurse will give him an injection to help with that.

d) Sure, I'll help sit him up so it is easier to breathe.

Mr Bartoli: My father is very breathless. Can you please help him?

1) Carer: _____.

Mr Bartoli: Why is my father having morphine?

2) Carer: _____.

Mr Bartoli: Can you do anything about his noisy breathing?

3) Carer: _____.

Mr Bartoli: Can you do something about his mouth and lips? They're very dry.

4) Carer: _____.

Unit 1: Transcripts of the videos

Video 1:

Carer: Good Morning, Mrs Peters.

Mrs Peters: Oh. Good Morning, dear. What is your name?

Carer: My name is Jane. I'm one of the carers here. You're new here, aren't you? I'll introduce you to the people at your table.

Mrs Peters: Thank you. I don't know anyone yet.

Carer: That's all right. I know you arrived yesterday. Mrs Peters, this is Mrs Perry on your left on the other side of the table.

Mrs Peters: Hello Mrs Perry. My name is Maria.

Mrs Perry: Pleased to meet you. My name is Susan.

Carer: And Mr French is on your left, next to you. Mr French, this is Mrs Peters.

Mr French: Hello. Nice to meet you. Please call me by my first name. I'm George. We are all very friendly here!

Mrs Peters: Thank you, George. You certainly are all very nice.

Video 2:

Mrs Peters: Hello, Jane. My daughter is here to visit me. This is Sarah, my daughter. Jane: Pleased to meet you.

Sarah: Nice to meet you too, Jane.

Mrs Peters: Where is the kitchen? I would like to make a cup of tea for my daughter.

Jane: That's nice. The kitchen is next to the front desk.

Sarah: Jane, where is the bathroom? I need to wash my hands.

Jane: It's on your right when you go out of the dining room.

Sarah: Thank you. Where is the sitting room? I have some cake to have with our tea.

Jane: The sitting room is over there on your left. There are two chairs by the window where you can sit and have your tea.

Mrs Peters: Thank you, Jane. That sounds lovely.

Answers

Activity 1:

Formal greeting	Informal greeting
My name is Jane. Pleased to meet you, Mrs Perry. Good Morning, Mrs Peters.	Hi, George. How are you today? Hello, I'm Jane. Hi, Susan. How are things? Nice to meet you, Maria.

Activity 2: 1c 2.b 3.a 4.c

Activity 3: Good Morning, Mrs Peters.
What is your name?
I'm one of the carers here.
This is Mrs Perry on your left.
Pleased to meet you.
Please call me by my first name.

Activity 4:
Carer: Good Morning, Mrs Peters.
Mrs Peters: Oh. Good Morning, dear. What is your name?
Carer: My name is Jane. I'm one of the carers here. You're new

here, aren't you? I'll introduce you to the people at your table.

Mrs Peters: Thank you. I don't know anyone yet.

Carer: That's all right. I know you arrived yesterday. Mrs Peters, this is Mrs Perry on your left on the other side of the table.

Mrs Peters: Hello Mrs Perry. My name is Maria.

Mrs Perry: Pleased to meet you. My name is Susan.

Carer: And Mr French is on your left, next to you. Mr French, this is Mrs Peters.

Mr French: Hello. Nice to meet you. Please call me by my first name. I'm George. We are all very friendly here!

Mrs Peters: Thank you, George. You certainly are all very nice.

Activity 5: 1.c 2.e 3.a 4.b 5.d

Activity 6: 1. annoyed   2. feel they are treated like children   3. full name   4. more

Activity 7:

bathroom

dining room

front desk

kitchen

laundry

linen room

sitting room

treatment room

Activity 8: 1. treatment 2. laundry 3. dining 4. linen 5. sitting 6. front desk 7. bathroom 8. kitchen

Activity 9:

Mrs Peters: Hello, Jane. My daughter is here to visit me. This is Sarah, my daughter. Jane: Pleased to meet you.

Sarah: Nice to meet you too, Jane.

Mrs Peters: Where is the kitchen? I would like to make a cup of tea for my daughter.

Jane: That's nice. The kitchen is next to the front desk.

Sarah: Jane, where is the bathroom? I need to wash my hands.

Jane: It's on your right when you go out of the dining room.

Sarah: Thank you. Where is the sitting room? I have some cake to have with our tea.

Jane: The sitting room is over there on your left. There are two chairs by the window where you can sit and have your tea.

Mrs Peters: Thank you, Jane. That sounds lovely.

Unit 2: Transcript Video 1:

Carer: Hello, Mrs Peters. Would you like a cup of tea?

Resident: Thank you, Sally. Yes, I would like a cup of tea.

Carer: Do you like your tea weak or strong?

Resident: I like it quite strong. I don't like weak tea at all.

Carer: OK. Would you like some milk in your tea?

Resident: Yes, please but only a little bit.

Carer: Sure. I'll just put a dash of milk in your tea.

Resident: Thank you. I don't like my tea when it is very milky.

Carer: Do you take sugar in your tea?

Resident: I know I shouldn't have any sugar but I like just a bit in my tea.

Carer: How much sugar would you like?

Resident: Oh, just half a teaspoon, please. That's enough sugar for me.

Carer: All right. There you are. Half a spoonful of sugar. Can you stir it yourself?

Resident: Yes, I can manage. Can I take a biscuit from the plate as well?

Carer: Of course you can!

Transcript video 2

Jane: Mr Georgio, it's time for breakfast now.

Mr Georgio: Breakfast? I've already had my breakfast.

Jane: Not today, Mr Georgio. You haven't had your breakfast yet today.

Mr Georgio: Yes, I have. It's time for lunch now.

Jane: It's 8 o'clock in the morning. I've come to take you to the dining room for breakfast.

Mr Georgio: Are you sure it's only 8 o'clock? I would prefer to have my lunch now.

Jane: Have a look at the clock, Mr Georgio. Can you see it on the wall? It's a few minutes after 8 in the morning.

Mr Georgio: Oh dear! I really thought that it was lunch-time.

Jane: That's OK. Come and sit at your table now. Some of the other residents are waiting for you.

Mr Georgio: Oh, well I suppose it's all right. I feel like a cup of tea. Is there any tea ready?

Jane: I'll go and have a look for you right now.

Mr Georgio: Thank you, dear. A cup of tea is just what I want now.

Unit 2: Answers

Activity 1: 1.e　　2.h　　3.a　　4.f
　　　　　　5.g　　6.c　　7.d　　8.b

Activity 2: 1.c　　2.b　　3.c　　4.c

Activity 3: e) a cup of tea?　　c) tea weak or strong?
a) dash of milk　d) half a teaspoon of sugar
b) biscuit from the plate

Activity 4: 1.soup　2. bread　3. tea　4. juice
　　5. ice-cream　6. yoghurt　7. cake　8. biscuits

Activity 5: 1.T　2.F (they have some sense of smell)
3.T　　4.T　　5.F (they get out of the habit of eating)

Activity 6: <u>break</u> - fast　　<u>morn</u> - ing tea
　　　　<u>lunch</u>　<u>aft</u>- er - noon <u>tea</u>
　　　　<u>din</u> - ner　　<u>sup</u> - per

Activity 7: 1.d　2.e　3.f　4.a　5.c　6.b

Activity 8: 1. *is*　2. *had*　3. *is going*　4. *thinks*　5. *feels*

Activity 9: 1.d　2.f　3.a　4.b　5.c　6.e

Activity 10: (1) breakfast　(2) today　(3) morning
　　　　　(4) lunch　(5) tea　(6) look

Unit 3: Transcript Video 1:

Carer: Hello Mrs Greene. Do you need some help?
Mrs Greene: Yes please Suzy.
Carer: What do you want me to do?
Mrs Greene: Will you take me to the toilet? I can't manage myself.
Carer: Sure. Do you want to go now?
Mrs Greene: Yes. I need to go now. Will you help me out of the chair first?
Carer: Yes, of course. Give me your arm and I'll help you up. That's right.
Mrs Greene: Thank you.
Carer: Now, stand up. There you are.
Mrs Greene: Can you take me to the toilet, please?
Carer: Yes. We'll go now. Do you want to go to the dining room afterwards?
Mrs Greene: No, I won't go to the dining room today. I don't feel like eating lunch.
Carer: OK but you should have a drink. It's very hot today.
Mrs Greene: Yes. I see. I'll have a glass of juice in my room later.
Carer: Good idea. I'll bring you a glass of apple juice. I'll put it on the table for you.

Transcript video 2

Jane: Mrs Walker, did you go to the toilet tonight?

Mrs Walker: Yes, I did but I had some problems with my bowels.

Jane: Oh. What was the bowel movement like?

Mrs Walker: I just passed a few hard lumps. It hurt a lot to go to the toilet.

Jane: I see. You take a laxative every night, don't you?

Mrs Walker: Yes. I take some Senna tablets every night. They usually work well. I don't know why I am having all these problems now.

Jane: You've been taking some painkillers lately for your back pain, haven't you? They can make you constipated.

Mrs Walker: Oh, I didn't know that. Can you give me something to help me go to the toilet?

Jane: Sure. I can give you some suppositories tonight. They will help make your stool softer and easier to pass.

Mr Walkers: All right, I don't like them much but constipation is very unpleasant.

Jane: No, it certainly isn't very comfortable. Try to drink a bit more water as well. That will help you too.

Mrs Walker: All right, I'll drink more water if it will help me.

Unit 3: Answers

Activity 1: 1.f 2.e 3.d 4.c 5.a 6.b

Activity 2:

1. Mrs Greene wants to go to the toilet.
2. Mrs Greene cannot manage herself. (or can't manage herself)
3. Mrs Greene asks the nurse to help her out of the chair.
4. The nurse asks Mrs Greene to stand up.

5. No, Mrs Greene does not want to go to the dining room after she goes to the toilet. (or doesn't want).

6. The nurse tells Mrs Greene to have a drink.

Activity 3: 1.d 2f .3.a 4.c 5.b 6.e

Activity 4: (a) (c) (d) (b) (f) (e)

Activity 5: 1.bowels   2. toilet   3.wee   4.stools   5.pooh

Activity 6: 1.c 2.c 3.a 4.b 5.b

Activity 7: 1. bowel chart   2. medicine   3. rectum
           4. loose   5. constipated

Activity 8: 1.c 2.f 3.e 4.b 5.d 6.a

Activity 9:

Resident: I had some problems with my bowels tonight.

Carer: I see. Are you constipated?

Resident: Yes. I just passed a few hard lumps and it hurt a lot.

Carer: I'll give you a laxative to make the stool softer and easier to pass.

Resident: Thank you. It is very uncomfortable.

Carer: It's important to drink a lot of water when you are constipated.

Resident: All right, I'll have a glass of water now.

Unit 4 Answers

Transcript Video 1:

Carer: Hello, Mrs Peters. Will you come for a walk in the garden with me? It's a lovely day today.

Mrs Peters: Yes, it is lovely today. It's quite sunny. But my knee is aching and I don't think I can walk far at all.

Carer: That's OK. We'll only have a short walk. It's important for you to have some exercise every day.

Mrs Peters: Yes, you're right. I try to get some exercise every day but today it's difficult to get moving.

Carer: Do you use a walking stick when you walk around?

Mrs Peters: Yes. My walking stick is over there, on the chair.

Carer: Ah yes. I can see it. I'll get it for you. Would you like me to rub some anti-inflammatory cream into your knee first?

Mrs Peters: No, thank you. I took a painkiller half an hour ago. I think I'll be all right.

Carer: That's good. I'll help you out of the chair now so you can stand up.

Mrs Peters: Thank you. If you help me to stand up, I'll use the walking stick to steady myself.

Carer: All right. One, two, stand. That's it. Are you steady?

Mrs Peters: Just a minute. Yes. I'm quite steady now.

Carer: Good. Take my arm and we'll go for a short walk outside.

Mrs Peters: Thank you, dear. That would be nice. I want to see the roses in the garden. They look beautiful from my window.

Video 2

Carer: Good Morning, Mr George. I'll take you to the dining room now. It's almost time for lunch.

Mr George: Oh, hello Sarah. I've been reading my newspaper this morning. I didn't realise that it's lunch time.

Carer: Yes. It's lunch time now. I've got your walking frame for you. I'll put it in front of you.

Mr George: Yes, thank you. I need the walking frame to walk. I'm quite unsteady on my feet.

Carer: Can you stand up and hold onto the walking frame facing me?

Mr George: Yes. I'm a bit slow standing up, I'm afraid. It's my arthritis.

Carer: That's all right. Take your time.

Mr George: Yes, I'll have to take my time. My joints are a bit stiff today. There we are.

Carer: That's good. Can you bend your elbows slightly and hold onto the handgrips?

Mr George: Yes. I'm ready now.

Carer: Can you put your frame a little in front of you and take a few steps towards me?

Mr George: Ooh, it is hard to get going this morning. I'll try to start walking now.

Carer: Mr George, try not to lean back or you might fall. It's better to lean forward a little.

Mr George: Oh. OK. Is that better?

Carer: Yes, that's better. Here's your chair now. Can you turn around? That's it. Try to sit down on the chair. It's right behind you.

Mr George: Can I sit now? I won't fall?

Carer: Yes. You can sit now.

Answers

Activity 1: 1. mobility scooter   2. walking frame   3. walking stick

Activity 2: 1. balance 2. arthritis   3. steady   4. mobility 5. vehicle

Activity 3: 1.b   2.c   3.c   4.b   5.a

Activity 4: e) walk in the garden   f) can't walk far

c) .use a walking stick?   d) anti-inflammatory cream on knee?

h) took a painkiller   g) help out of chair?

 a) are you steady?   b) see roses in garden

Activity 5: (1) walk around . (2) rub (3) took (4) stand up (5) use (6) take

Activity 6: 1.F 2. T 3.T 4.T 5.F 6.T

Activity 7: 1.c 2.f 3.e 4.a 5.b 6.d

Activity 8: 1.a 2.c 3.c 4.b 5.c

Activity 9:

1. Can you stand up facing me?
2. Can you bend your elbows slightly?
3. Can you put your frame in front of you?
4. Try not to lean back or you might fall.
5. Can you turn around?
6. Try to sit down on the chair.

Activity 10: (1) need (2) hold onto (3) take (4) bend (5) put (6) lean back (7) lean forward (8) turn around

Photo credits.

pg x   Photo by Phasmatisnoxat en.wikipedia

http://upload.wikimedia.org/wikipedia/commons/thumb/0/04/Cane_IMG_2994.jpg/80px-Cane_IMG_2994.jpg?uselang=en-gb

Unit 5 answers:

Transcripts. Video 1

Carer 1: Mrs Summers, Heather and I are going to turn you on your side for a while. Is that OK?

Resident: Oh no. Leave me on my back. I don't want to go on my side.

Carer 1: I'm sorry, Mrs Summers. It's very important to change your position when you are in bed for a long time.

Resident: But you moved me a few minutes ago.

Carer 1: I know it seems like a short time, but it was actually more than two hours ago.

Carer 2: We need to check your hip now as well. It was a little red last time we turned you.

Resident: I suppose so. Will I have to stay on my side for a long time?

Carer 1: No, not too long. Can you stay on your side until dinner time? Before dinner, we'll roll you onto your back again and sit you up for your dinner.

Resident: You won't forget me, will you?

Carer 1: No, we won't forget you. We'll come back in time for you to get ready for your dinner.

Carer 2: All right. We'll help you roll onto your right

side now. Can you put your arm around Jenny and I'll turn you onto your side?

Resident: Oh dear. Yes, I can hold onto Jenny...um, oh..

Carer 1: That's the way. Now, Heather is going to have a look at your hip to check the skin.

Carer 2: Let me see. The skin is still a little red but it's not broken.

Resident: What do you mean, broken? I haven't broken my hip, have I?

Carer 2: Oh no, that's not what I mean! We check to see if there is a break in your skin. Even a small tear in the skin can cause a nasty sore to form. That's what we call a pressure ulcer.

Carer 1: But your skin isn't broken at all. It's just a little red. Because the skin is a little red it's important to keep changing your position every few hours. It takes the pressure off your skin so pressure ulcers don't form.

Resident: What about my things? I can't reach them on this side.

Carer 2: That's OK. I'll put your bedside table on the other side for you. Here's your drink and your magazine.

Is there anything else that you need?

Resident: Oh yes, where is my box of tissues, dear?

Carer 1: Here it is. I'll put it next to the magazine.

Resident: Thank you, dear.

Transcript video 2

Nurse: Can we go through Mrs Summer's Waterlow Chart now? I need to update her details.

Carer: Sure.

Nurse: Right. What does she weigh now?

Carer: Ah, Let me see. I weighed her this morning and she is only 56 kilos.

Nurse: So, she is below average weight. That's three. Can you tell me what her skin is like?

Carer: Her skin is quite dry. I put a moisturising cream on every morning.

Nurse: OK. That's one. She's female so that's two. What about her age? How old is Mrs Summers?

Carer: Ah, she's 86. She's over 81 so it's a 5.

Nurse: Her appetite is OK at the moment, I have been watching her over the past few days.

Carer: All right. A zero for appetite.

Nurse: Does Mrs Summers have any problems with incontinence?

Carer: Yes, she's incontinent at times. She wears a pad during the day and at night.

Nurse: All right, so one for incontinence. I know she uses a walking stick to mobilise so that's a three.

Carer: I don't know what medication she takes so I can't help you with the next section.

Nurse: That's OK. I'll just check her drug chart. Oh yes, she takes steroids so I'll circle number four.

Carer: Now just the last section. I know she has diabetes so we circle number 4. What about the other ones?

Nurse: Well she's not a smoker but she does have poor circulation. We need to circle 5 as well.

Carer: That's all, isn't it?

Nurse: Yes. I just have to add the numbers to work out her risk. 3,1,2.5.0,1,3,4,4,5. That's 28. That's a high risk for pressure ulcers. Can you please make sure that you change her position frequently when she is in bed or in a chair?

Carer: Sure.

Nurse: Can you put an air mattress on her bed too, please?

Carer: OK. I'll put it on this afternoon.

Nurse: That's great. The last thing is to check that she is kept dry. We need to keep an eye on her skin condition too.

Carer: No problem. I'll keep an eye on it.

Answers

Activity 1: 1.e   2.c   3.d   4.b   5.a

Activity 2: 1. The carers are going to turn Mrs Summers on her side.

2. The carers need to check Mrs Summer's hip because it was red.

3. They will roll Mrs Summers onto her back again and sit her up for dinner.

4. No, Mrs Summers does not have any broken areas but the skin over her hip is red.

5. The carer puts Mrs Summers' drink, magazine and box of tissues within Mrs Summers' reach.

Activity 3: (c)   (b)   (e)   (a)   (f)   (d)

Activity 4: 1.d   2f   .3.a   4.c   5.b   6.e

Activity 5: 1.b   2.a   3.c   4.b   5.b

Activity 6: 1.b   2.a   3.a   4.c   5.a

Activity 7: 3 - below average, 1 - dry, 2 - female, 5 - 81+, 0 - average appetite,1 - incontinent of urine, 3 - uses walking aid, 4- steroids, 4-diabetes, 5- poor circulation. Risk 28 - High

Activity 8: 1.score   2.skin   3.bedbound   4.medications   5.turned

Activity 9: 1.d   2.f   3.a   4.e   5.c   6.b

Activity 10: 1.update   2.weigh   3.above average   4.healthy   5.incontinence   6.steroids   7.four   8.smoker

Unit 6 answers:

Transcripts. Video 1

[Sound of a resident calling out] Help! Help!

Carer : Mrs Mitchell, are you all right?

Resident: Oh help. I'm over here.

Carer : Oh dear. You've had a fall. Don't try to get up just now. Stay where you are and I'll get someone to help me get you up.

Resident: Ooh. I fell over. It was silly of me. I'm sorry.

Carer 1: That's all right. I'll help you now. Can you tell me what happened?

Resident: I needed to go to the toilet. It's those water tablets I take... the ones for my heart. I was worried that I was going to wet myself. It was still dark and I couldn't see very well.

Carer 1: Oh, yes. It's very difficult when you take that type of medicine, isn't it? That's why it's a good idea to call us so we can help you to the toilet at night.

Resident: I suppose so. I just didn't want to bother you. You carers always seem so busy.

Carer 1: Oh please don't worry about that. It's more

important that we help you so you don't have an accident. We don't want you to injure yourself.

Resident: I see. Can I get up now? Oh dear. I'm really sorry.... I think I've wet the floor. That's awful for you. I'm so sorry..it's very embarrassing. I don't usually wet myself.

Carer 1: That's OK. You said you were taking medicine which makes you go to the toilet a lot. It's not your fault at all. I'll help freshen you up in a minute and you'll feel better. I just need to call the nurse to check that you are OK to stand up...you know, check that you haven't broken anything. Is that all right with you?

Resident: Oh dear. I don't think I've broken anything. I hope not.

Carer 1: It looks as if you are OK but I still need to check with the nurse first. I'll press the call button now. She shouldn't be too long. [buzzer]

Resident: All right, dear. I won't try to move. I'll wait for the nurse to come.

Transcript video 2

Nurse: How is Mr Field today? Has he had any falls?

Carer: He's OK today but he did have a fall last night. Fortunately, it was only a small fall.

Nurse: Oh dear. That's no good. What happened?

Carer: He took a laxative yesterday and was worried that he couldn't get to the toilet in time.

Nurse: That must have been difficult for him. He takes a lot of care with his personal hygiene.

Carer: Yes, he does. Unfortunately, he woke up in the middle of the night and couldn't see very well.

Nurse: Was his night light put on last night?

Carer: No. I don't think it was.

Nurse: I see. Well, it's very important that his night light is left on every night as his vision is quite poor. I'll make sure that I ask the night staff to check his night light tonight.

Carer: I think that he is still wearing his old slippers at night. They are not safe at all.

Nurse: OK. I'll have a chat with him and explain that his slippers may make him trip over. What about his call bell? Does he use it all the time?

Carer: I don't think he likes it much. He often seems to leave it on his locker.

Nurse: I can understand that he finds it a nuisance but it's important that he keeps it with him at all times. He's quite unsteady on his feet and needs a bit of help. Could you talk to him about it? You seem to have a good rapport with him.

Carer: Yes, I get on well with him. I might be able to explain how important it is to ask for help.

Nurse: That would be great. I don't want him to have a bad fall.

Carer: Neither do I!

Unit 6: Answers

Activity 1: 1.d   2.a 3.e 4.c   5.b

Activity 2: 1.b 2.c 3.a 4.c   5.b

Activity 3: 1.e 2.a 3.b 4.c 5.d

Activity 4: 1. fell over 2. wet myself 3. mean 4. bother 5. feel   6. hurt myself

Activity 5: 1.c 2.b 3.a 4.c 5.a   6.b 7.a 8.c

Activity 6: 1. wear 2. unsteady 3. surroundings 4. toilet 5. incontinent 6. drowsy

Activity 7: 1.c 2.b 3.c 4.a 5.b

Activity 8:

FALL RISK FACTOR CHECKLIST Resident: Mr George Friend Room 1		Yes / No
Vision	poor vision / wears glasses	Yes
Mobility	unsteady gait / poor balance / uses mobility aid / unsafe shoewear	Yes
Behaviour	confused at times / disoriented at night	Yes
Continence	incontinent of urine / incontinent of faeces / uses laxatives	Yes
Medication	painkillers / sleeping tablets / diuretics (water tablets)	No
FALL RISK	LOW MEDIUM HIGH **Important:** If resident has a HIGH falls risk, Commence Fall Risk Reduction (night light, frequent toileting, call bell within reach, keep environment clear at all times)	**HIGH**

The nurse suggests a night light, not wear old slippers, use call bell at all times

Activity 9: 1. Carer: Not today, but he did have a fall last night.

2. Carer: No, I don't think it was.

3. **Carer:** I don't think he likes it much. He often seems to leave it on the locker.

4. **Carer:** Yes, I might be able to explain how important it is to ask for help.

Activity 10 :

Carer: Hello, Mr Field. I need to make sure that your night light is switched on.

Resident: Why do I need the light on?.

Carer: It's so you can see if you wake up in the night and need to go to the toilet.

Resident: Oh, yes. That's a good idea.

Carer: Mr Friend, it would be a good idea if you didn't wear your old slippers any more.

Resident: But they are so comfortable!

Carer: I know but you might trip over easily if you wear them.

Resident: Well, my daughter brought me some new ones so I guess I can wear those.

Carer: Good idea. One last thing - could you use the call buzzer if you want to get up at night?

Resident: I don't like to ring late at night.

Unit 7 answers:

Transcripts. Video 1

Carer : Shauna, can I talk to you about Mrs Littlewood. She's really confused all of a sudden.

Nurse: Mrs Littlewood? Which room is she in?

Carer: Jessie Littlewood. She's in room 18, next to Mrs Jones.

Nurse: OK. Yes, I know her. What's the problem?

Carer: Well, she's not making sense at all. I asked her if she wanted some lunch and she became quite aggressive. She told me to get out of her room.

Nurse: I see. She's usually quite calm.

Carer: Yes. That's why I was a bit worried. And the other thing is that she seems to be seeing things..you know, hallucinating.

Nurse: What, you mean, she's seeing things that aren't there?

Carer: Yeah, that's right. She keeps saying that her mother is having a cup of tea with her and that I should leave them alone. But her mother isn't there obviously.

Nurse: Right. I checked her urine yesterday and it was OK. What about her bowels? Is she constipated? Sometimes constipation can make elderly residents confused.

Carer: No, she hasn't had any problems with her bowels. She has her bowels open every day.

Nurse: All right. There is another cause of confusion in the elderly. Sometimes a change in medication causes problems. Let me check Mrs Littlewood's drug chart... oh yes, that's right. The doctor changed her painkillers this week. She started taking a painkiller with codeine two days ago.

Carer: Oh yes, that was because she hurt her knee when she fell over.

Nurse: Yes, but I think she's had a bad reaction to the codeine in the painkiller. I won't give her any more codeine and I'll call the doctor to review her medication.

Carer: OK, thanks. I'll go and check her again to make sure she is all right.

Video 2

Carer : Hello Mr Browne, how are you today?

Resident: Who are you? I don't know you. Do you work here?

Carer : Yes, Mr Browne. I work here. My name is Rosemary. I'm one of the carers.

Resident: I can't see you very well.. who are you? I don't have my glasses.

Carer: It's Rosemary, Mr Browne. There are your glasses. You've got them on. Can you see me OK?

Resident: Oh, yes. That's better. Where's the door? I have to go to work.

Carer: Oh but it's night-time now. It's not time to go out now. You've just had your dinner with the other residents.

Resident: No, no, no. I have to go to work now. Look. Where's the door? Tell me where the door is now. I need to go.

Carer: It's a bit dark outside now. Remember that you only work in the day-time. It's time to get ready for bed soon. We're going to have some supper in a minute. Would you like a warm drink before bed?

Resident: No, I haven't got time for a drink. I have to go to work.

Carer: OK, I see. What about having a small drink first? You still have time. Let's go into the dining room and I'll get you some warm milk.

Resident: Yes, I'm quite thirsty. I need my bag first. I'll have to go out soon.

Carer: Sure. That's your bag, isn't it? I'll bring it into the dining room for you.

Resident: Yes, I really must have my bag. I have to go to work later. But I can't find where my glasses are.

Carer: You've got them on! I gave you your glasses a minute ago.

Resident: Oh yes. So you did. I can see you well now.

Unit 7: Answers

Activity 1: 1. serious 2. acute 3. confused 4. active 5. tired 6. real

Activity 2: 1.b 2.a 3.c 4.b 5.a

Activity 3: 1. confused 2. real 3. delirium 4. drugs 5. dementia

Activity 4: 1.d 2.e 3.a 4.b 5.c

Activity 5: 1.b 2.c 3.a 4.b 5.c

Activity 6: 1.b 2.a 3.c 4.c 5.a

Activity 7: 1. carers 2. dinner 3. day-time 4. drink 5. dining room

Activity 8: 1.T 2.F 3.T 4.F 5.F

Activity 9:

1. Mr Browne has become very confused.
2. The carer diverts Mr Browne's attention.
3. Mr Browne says things which don't make sense.
4. Mr Browne wanders a lot lately.
5. Mr Browne doesn't recognise the carer at all.

Activity 10: (1) d (2) e (3) c (4) b (5) a

Unit 8 Transcripts. Video 1

Carer: Hello, Mrs Grey. Have you taken all your medication this morning?

Resident: Oh hello, Ann. Yes. I think I've had all of them. I have them with me in my room so I can take them myself.

Carer: Yes, I know. Your chart says that you self-medicate. Do you mind if I check against your drug chart to make sure?

Resident: No, I don't mind. It's always a good idea to check.

Carer: Let me see. Did you take Furosemide 40 mg and Omeprazole 20 mg?

Resident: Yes. I took the Furosemide, that's the water tablet. I don't like it much but I know it's important for my heart.

Carer: Yes. That's right. What about the other one?

Resident: The Omeprazole? Yes. I had that. It's a capsule. It helps line my stomach so I won't get an ulcer. At least that's what the doctor said.

Carer: Great, well that seems to be correct.

Resident: Can I ask you about one of my tablets which I am having trouble with?

Carer: Yes, sure. Which one do you mean?

Resident: It's the one that I have at lunch-time. It's very large and I can't swallow it easily. It gets stuck in my throat.

Carer: Right. I'll check with the nurse but I think you could have an elixir instead.

Resident: I see. You mean that I could take a liquid instead of the hard tablet?

Carer: Yes. It may be easier for you.

Resident: Thanks, it would be much better to do that. There is something else I need too. Could I have a painkiller this morning, please? I have a lot of pain in my hip today.

Carer: That's no good. I'll just check your chart to see what your 'as required' medication is. Let me see. Yes, I can give you some painkillers for your hip. I'll get them now.

Resident: Thank you, dear. I don't need the as required medication much but I certainly need it today.

Video 2

Nurse: Julie, do you have a minute? I need to fill you in about Mr George's best interest meeting.

Carer : OK. I'm free at the moment. Why did you have the meeting?

Nurse: We had to have the meeting to discuss how to give Mr George his tablets.

Carer: Right. He has become a lot more confused lately, hasn't he? He doesn't like having a shower or eating breakfast any more. He becomes very upset.

Nurse: Yes. Unfortunately, he is more confused now and more anxious too. When I try to give him his blood pressure tablets he thinks I am trying to poison him.

Carer: I see. That's why he doesn't want to take them. Does he have to have the tablets?

Nurse: Yes. It's very important that he has the tablets to keep his blood pressure under control.

Carer: How can we give him the tablets if he doesn't want to take them?

Nurse: Well, that's what the meeting was about. His daughter came in to talk with us..with the doctor, the pharmacist and I. I explained to Mr George's daughter that we had assessed his mental state again recently and had noticed that he is becoming very confused. The pharmacist explained that Mr George should continue to take his tablets.

Carer: And, she understood?

Nurse: Yes, she knows that we need to give him the medication and has given us permission to hide his medication in a glass of milk or in some custard so he doesn't know he is taking it.

Carer: So, it's OK to do that?

Nurse: Yes. It's OK because we have written permission and because Mr George is no longer able to make the important decision himself. I have changed his care plan to say that his tablets must be crushed and given to him in milk or custard. He won't know that he is taking the tablet so he won't get upset about it.

Carer: How will I crush the tablet?

Nurse: I'll give you a mortar and pestle to use. The mortar is a little bowl and the pestle is a small implement which is used to grind the tablet to a powder. I'll leave it in the treatment room.

Carer: OK, thanks.

Unit 8: Answers

Activity 1: 1.d 2.f 3.e 4.a 5.c 6.b

Activity 2: 1. drug errors 2. swallow 3. milligrams
             4. mls   5. prn   6. pharmacy

Activity 3: 1.c 2.b 3.a 4.b 5.c

Activity 4: 1.c 2.e 3.a 4.b 5.d

Activity 5: 1.a 2.b 3.c 4.b

Activity 6: 1. tablets 2. poison 3. pharmacist
             4. permission 5. crush

Activity 7: 1.d 2.a 3.c 4.b

Activity 8:

Nurse: Can you please crush Mr George's tablets before you give them to him?

Carer: OK. How do I do that?

Nurse: Use the mortar and pestle in the treatment room to help you.

Carer: How should I give him the crushed tablets? He always refuses to take his tablets.

Nurse: You can put the crushed tablets in some custard so he doesn't know he's taking them.

Carer: Am I allowed to do that?

Nurse: Yes. There is a consent form in his care plan signed by his daughter as he is too confused to make the decision himself.

Activity 9: 1.F 2.T 3.T 4.F

Activity 10:

1. Mrs Simpson takes a water tablet twice a day.
2. Omeprazole is a capsule to prevent stomach problems.
3. Senna tablets are laxatives which help with constipation.
4. Rheumatoid Arthritis is treated with Methotrexate tablets once a week.

Unit 9 Transcripts. Video 1

Carer: How are you feeling this morning, Mrs Cotteral?

Resident: Oh hello, Judy. I'm, ah..I'm all right.

Carer: Are you sure? Your hip was painful yesterday, wasn't it?

Resident: Yes, it was hurting a bit. I didn't want to say anything. I can manage..

Carer: Oh no, Mrs Cotteral. You don't have to put up with the pain. Can you tell me about the pain level? Is the pain stopping you from doing your normal activities? Like walking around, washing yourself and getting dressed?

Resident: Yes. The pain does make it hard to walk around. But it's only my right hip. I don't like to complain about it.

Carer: That's OK. It's not complaining. We like to know if you are in pain because there are some things we can give you to relieve the pain.

Resident: I'm just worried about taking the tablets for pain. They're addictive you know and I don't want to depend on them every day.

Carer: I understand that you are worried about it, but there are some painkillers that are not addictive. And also, it's important to be able to keep your independence.

Resident: I suppose you are right. I don't like not being able to walk around easily. You know I like to walk in the garden every day.

Carer: Yes, I know you love the garden. That's why I thought your hip might be aching today. It's the first time I haven't seen you in the garden for a long time.

Resident: I didn't want to make a fuss. Maybe the pain will go away soon.

Carer: Would you be willing to try some pain relieving gel?

Resident: What's that?

Carer: It's an anti-inflammatory gel which I rub into your hip. It helps stop the aching.

Resident: Well, it sounds like a good idea. Yes, I'd like to try the gel, if it's all right.

Carer: Sure. I'll get it right now.

Video 2

Carer : Hello Mr Sainsbury. How are you feeling?

Resident: Oh, Janet. It's so painful. I know they said I was ready to leave hospital but I really don't think they know how painful it is.

Carer: Oh dear. I understand that it must be difficult for you. A shoulder operation can be very painful. Can you tell me what the pain's like at the moment?

Resident: Yes. It's very painful. It's still very painful. I don't think I'm ever going to be all right again.

Carer: I can see how you might think that. Unfortunately, shoulder operations take a long time to heal. Can you tell me how painful it is on the pain scale? You remember, zero is no pain and 10 is the worst pain you can imagine.

Resident: Yes, I remember the pain scale. They use it all the time in hospital. It's a seven this morning..no,actually it's an eight. I never know what the difference is.

Carer: Well, if the pain stops you from moving around then it's severe pain. It's different for everyone but severe pain is anything between 7 and 10. What does the pain feel like?

Resident: It seems to ache most of the time. It's a really sharp pain when I try to lift my arm.

Carer: Are you able to do the exercises the physiotherapist gave you?

Resident: No, I can't do them. I try to but I can't manage. It's too painful.

Carer: Are you taking your painkillers before you do the exercises?

Resident: No. I don't want to take too many painkillers. It's not good for you.

Carer: I understand your concerns but, the thing is that shoulder operations are very painful. But, it's also very important that you do your exercises so your shoulder doesn't become too stiff.

Resident: Oh yes. I think that the physiotherapist mentioned that to me. I really try to do them but it hurts a lot.

Carer: Yes, it is quite difficult at first. That's why it's important not to miss your painkiller doses. If you have the painkillers regularly, it keeps the pain under control much better. You have some painkillers due now. If you take them and wait for 20 minutes, you'll be able to do your exercises.

Resident: All right. I'll try that.

Unit 9: Answers

Activity 1: 1.d 2.c 3.e 4.a 5.b

Activity 2: 1.b 2.a 3.c 4.a 5.b

Activity 3: 1. painful 2. put up with 3. walk around
4. relieve 5. addictive 6. rub

Activity 4: 1.d 2.a 3.b 4.e 5.c

Activity 5: 1. lasts 2. aching 3. independent
4. addicted 5. dementia

Activity 6: 1.c 2.e 3.a 4.b 5.d

Activity 7: (1) It's still very painful. (2) it's an eight. (3) It's a really sharp pain (4) I try to but I can't manage (5) I don't want to take too many painkillers.

Activity 8:

What pain like?

How painful on pain scale?

Sharp pain when lift arm

Able to do exercises?

Take painkillers before exercise

Important not miss dose

Painkillers due now

Wait 20 minutes, able to do exercises

Activity 9: 1.T 2.T 3.F 4.F 5.F 6.T 7.F 8.F

Unit 10 Transcripts. Video 1

Carer: Mrs Downing, is it all right if I have a chat with you about your advance care plan?

Resident: Oh hello, Sally. Certainly. What do you want to talk about?

Carer: Well, we have to make sure that your advance care plan is up-to-date. You know, if your condition gets worse.

Resident: Oh yes, I have been thinking about it a lot. This disease is awful. I know I'm getting weaker and weaker.

Carer: I know it must be difficult for you to talk about end of life care but it's important that we understand what your wishes are.

Resident: It's all right, dear. It is a little hard to think about it but I would rather sort things out now..you know, while I'm still able to make my own decisions.

Carer: That's good! The first thing I'd like to know is what type of care you would like towards the end of your life.

Resident: I would like to be kept comfortable at the end. I just want medication which will help with the pain. I know that there isn't any treatment for my condition but I am afraid of being in pain.

Carer: It's quite normal to worry about that but there are very good ways now to control pain. You'll have a special pump which will give you a small dose of painkillers over the whole day. The pump works very well.

Resident: That's a relief. I don't want any tubes left in either. I want to look as normal as possible. I'm worried about how my family will cope.

Carer: OK. I'll tick the box that says you don't want any tubes or drips unless they are used to make you more comfortable.

Resident: Can my family stay with me at the end? My son and daughter have asked if they can stay. I don't know if that's allowed.

Carer: Yes, of course it's all right. We can make up a bed in your room for your family and we'll order a meal for them while they are staying with you.

Resident: Oh, that is kind.

Carer: Is there anything that you'd like to have in your room? Some people like to have some soft music on or they put up some photographs, for example.

Resident: Do you know, I prefer jazz to soft music. Would it be all right to play my jazz CDs? I already have photographs of my family all around me now. They are very important to me.

Carer: We can play any music you like! We can make a playlist of your favourites later if you wish.

Resident: Actually, I've been doing that with my grand-daughter. She's very good at it.

Carer: That's great! There's one last question, Mrs Downing. I know you enjoy Father Martin's service on Sundays. Would you like Father Martin to come and talk to you at any time?

Resident: Yes, I'd like that. He's such a nice young man. I find his words very comforting.

Carer: Yes, he is very nice. I'll phone him and ask him to have a chat with you this week. OK, your advance care plan has been updated now so I'll put it in your record. If you decide that you would like to change anything, just let us know.

Resident: Thank you, dear. I'll do that.

Video 2

Carer 1: Hello, we've come to make your father more comfortable.

Son: Thank you. I just got here to see him. Dad's been having a bit of trouble breathing.... I don't know if you can do anything for him.

   Carer 1: Yes, we can certainly make him a more comfortable. Mr Bartoli, Gemma and I are going to sit you up a bit more so you can breathe more easily.

   Son: I don't think there's any point trying to talk to him. He hasn't talked to me at all since I've been here.

Carer 2: Actually, it's a good idea to keep talking to your father even if you think he doesn't seem to be aware that you're here.

Son: Oh really? OK, I'll keep talking to him then. He seems very agitated. Is that normal? He's been trying to push off his blanket even though he feels very cold.

Carer 1: I see. I'll talk to the nurse about it. She can give him another injection to calm him down.

Son: OK but is it safe? What if he stops breathing?

Carer2: I can understand why you are concerned. It's very difficult to watch your loved one struggling to breathe. But, people who are in the last stages of heart failure like your father often have trouble breathing. It makes them very distressed.

Son: Yes, that's what's been happening with Dad. I don't want him to suffer but I'm worried about him having too much medication. The nurse said she was giving him some morphine as well. That can't be right, can it? I mean, I thought that morphine makes you breathe a lot slower.

Carer 1: You're right about morphine. It usually makes the breathing rate slower. Don't be concerned because your father is only getting a very small dose of morphine. It's just enough to help with the breathlessness.

Son: I see. I didn't realise. My sister is worried about...um, is it called the death rattle? She said she doesn't want to visit in case my father starts breathing like that.

Carer 2: It's OK. A lot of people worry about that. There is an injection the nurse can give your father to stop him having the noisy breathing you call the death rattle. She said that she can give him an injection as often as he needs it.

Son: All right. I'll tell my sister. Is there anything I can do for my father?

Carer 1: Just let him know you are here. Talk to him about all the things you think he should hear from you. It may be your last chance to talk to him.

Son: OK. How long do you think he has?

Carer 1: No-one can tell you that for sure, I'm afraid. The only thing I can tell you is that he is no longer talking to us and he is very breathless.

Carer 2: We'll keep changing his position and doing his mouth care to keep him comfortable. His mouth and lips are very dry because of the breathlessness.

Son: Thank you for explaining everything to me. I'll stay here with him for a little bit longer, if that's OK?

Carer 1: Of course. Stay as long as you like.

Unit 10: Answers

Activity 1: 1.c 2.d 3.e 4.a 5.b

Activity 2: 1.a 2.a 3.b 4.c 5.a 6.c

Activity 3: 1. I know it must be difficult for you to talk about end of life care.

3. It's quite normal to worry about that ...

Activity 4: 1. cared for 2. unnecessary 3. heart attack
4. depression 5. spiritual

Activity 5: 1. doctor 2. daughter 3. does not want
4. reviewed 5. nurse

Activity 6:

Dad trouble breathing

keep talking to your father

seems very agitated

injection to calm him down

worried - too much medication

very small dose morphine - help with breathlessness

worried about death rattle

injection stop noisy breathing

keep changing position and doing mouth care

Activity 7: 1.F 2.T 3.T 4.F 5.T 6.F

Activity 8: 1. c   2.a   3.d   4.b

Activity 9: 1.d   2.a   3.e   4.b   5.c

Activity 10: 1.d   2.a   3.c   4.b

www.ingramcontent.com/pod-product-compliance
Lightning Source LLC
Chambersburg PA
CBHW070231180526
45158CB00001BA/343